Julia Keller

Diagnosing the Downstream Impact of Extratropical Transition Using Multimodel Operational Ensemble Prediction Systems

Wissenschaftliche Berichte des Instituts für Meteorologie und
Klimaforschung des Karlsruher Instituts für Technologie
Band 57

Herausgeber: Prof. Dr. Ch. Kottmeier
Institut für Meteorologie und Klimaforschung
am Karlsruher Institut für Technologie
Kaiserstr. 12, 76128 Karlsruhe

Eine Übersicht über alle bisher in dieser Schriftenreihe erschienenen Bände
finden Sie am Ende des Buchs.

Diagnosing the Downstream Impact of Extratropical Transition Using Multimodel Operational Ensemble Prediction Systems

by
Julia Keller

Dissertation, Karlsruher Institut für Technologie
Fakultät für Physik
Tag der mündlichen Prüfung: 13. Januar 2012
Referenten: Prof. Dr. Sarah Jones, Prof. Dr. Christoph Kottmeier

Impressum

Karlsruher Institut für Technologie (KIT)
KIT Scientific Publishing
Straße am Forum 2
D-76131 Karlsruhe
www.ksp.kit.edu

KIT – Universität des Landes Baden-Württemberg und nationales
Forschungszentrum in der Helmholtz-Gemeinschaft

KIT Scientific Publishing 2013
Print on Demand

ISSN 0179-5619
ISBN 978-3-86644-984-8

Abstract

The extratropical transition of tropical cyclones (ET) often leads to a reduction in predictability for the synoptic development of the cyclones themselves, as well as for downstream regions. The improvement of numerical weather predictions for ET events requires a deeper understanding of the underlying physical and dynamical processes during the interaction of the tropical cyclone with the mid-latitude flow. Furthermore, new alternative approaches are desired to better account for the inherent uncertainty during ET using already existing forecast tools. In this study the predictability during ET events is examined in a new multimodel ensemble data set. Furthermore, the interaction between the transitioning tropical cyclone and the mid-latitude flow is explored in an kinetic energy framework.

Recent studies investigated the predictability during ET events based on the variability among members of single operational medium-range ensemble forecasts. The new THORPEX Interactive Grand Global Ensemble (TIGGE) combines several ensemble prediction systems (EPS) to a new multimodel ensemble. This new data set is used in this study to compare predictability during ET events in the different contributing EPS and to explore the potential benefits of the multimodel approach. TIGGE multimodel forecasts for five ET events are investigated using an empirical orthogonal function (EOF) analysis followed by a subsequent fuzzy clustering, enabling the extraction of the main possible development scenarios. In the present study, the characteristics of the TIGGE data set during the application of the described analysis technique are explored. It is examined, whether the TIGGE data set shows more variability and thus more possible development scenarios during ET events than the stand-alone ensemble forecast from the European Centre for Medium-Range Weather Forecast (ECMWF). Another important aspect of the study is to investigate, whether the individual EPS contribute differently to the development scenarios. It is shown that the individual EPS differ in their spread and their contributions to the distinct development scenarios. While some of the individual EPS are generally confined to only a few scenarios, other EPS contribute regularly to almost all scenarios. In principle, TIGGE contains more development scenarios than the ECMWF EPS. However, the full range of development scenarios is only found with the ECMWF included in the multimodel EPS.

Ensemble forecasts provide possible development scenarios, which may differ strongly during episodes with decreased predictability, like ET events. These forecast scenarios are used in this study to further investigate the underlying physical and dynamical processes causing the distinct developments. Therefore, an analysis of the eddy kinetic energy budget is applied to forecast scenarios, which are extracted from ECMWF ensemble forecasts for the ET of two tropical cyclones. A comparison of the eddy kinetic energy budget for these scenarios reveals the contribution of the transitioning tropical cyclone to the distinct developments and indicates the relevant processes during ET. It is shown that the duration of baroclinic conversion of eddy kinetic energy and the phasing between the transitioning tropical cyclone strongly influence the impact of the cyclone on the mid-latitude flow. Furthermore, the mid-latitude flow is sensitive to the interaction with a tropical cyclone only during a short time period, thereafter the impact of the albeit strong cyclone on the mid-latitude wave pattern decreases.

This study reveals new aspects of predictability and the physical processes during ET and points out future directions for further studies in this field.

Kurzfassung

Die außertropische Umwandlung von tropischen Wirbelstürmen (extratropical transition - ET) ist oftmals von einer deutlichen Verschlechterung der Vorhersagbarkeit für die synoptische Entwicklung der Wirbelstürme selbst, sowie für weiter stromabwärts liegende Regionen begleitet. Eine Weiterentwicklung der numerischen Wettervorhersage während ET erfordert ein verbessertes Verständnis der physikalischen und dynamischen Prozesse, die die Wechselwirkung des sich umwandelnden Sturms mit der Strömung der mittleren Breiten bestimmen. Weiterhin besteht Bedarf an der Entwicklung neuer Wege, die es ermöglichen, die Vorhersageunsicherheit während ET mit bereits existierenden Vorhersageprodukten noch besser zu erfassen. In der vorliegenden Arbeit wird zunächst die Vorhersagbarkeit während ET in einem neuen Multimodell-Ensemblevorhersagesystem untersucht. Im zweiten Teil der Arbeit wird die Wechselwirkung zwischen einem sich umwandelnden tropischen Wirbelsturm und der Strömung der mittleren Breiten anhand eines kinetischen Energie-Budgets näher analysiert.

Bisherige Studien konzentrierten sich auf die Untersuchung des Einflusses von ET auf die Vorhersagbarkeit für stromabwärts gelegene Regionen anhand der Vorhersageunsicherheit in einzelnen operationellen mittelfristigen Ensemblevorhersagen. Das neue "THORPEX Interactive Grand Global Ensemble (TIGGE)" vereint mehrere dieser einzelnen Ensemblevorhersagesysteme zu einem neuen Multimodell-Ensemble. Dieser neue Datensatz ermöglicht einen Vergleich der Vorhersagbarkeit während ET in den einzelnen zu TIGGE beitragenden Ensemblevorhersagesystemen. Zudem können

die möglichen Vorteile einer Multimodell-Ensemblevorhersage für ET er-
örtert werden. In der vorliegenden Arbeit werden zehn TIGGE Vorhersagen für fünf ET-Fälle mittels einer Bestimmung der Empirischen Ortho-
gonalfunktionen und einem anschliessenden Fuzzy-Clustering der Haupt-
komponenten ausgewertet. Diese Analysemethode ermöglicht es, die do-
minierenden Vorhersageszenarien aus der Ensemblevorhersage zu extra-
hieren. Ziel der Arbeit ist es, das Verhalten des neuen Datensatzes während der Anwendung der beschriebenen Analysemethode zu untersuchen
und die Vorhersagevariabilität in TIGGE während ET zu charakterisieren.
Es wird untersucht, ob TIGGE mehr Vorhersagevariabilität und daher mehr
mögliche Vorhersageszenarien als das einzelne Ensemblevorhersagesystem
des Europäischen Zentrums für mittelfristige Wettervorhersage (EZMW)
zeigt und damit die tatsächliche Entwicklung öfter einschliesst. Ein wei-
terer wichtiger Aspekt dieser Studie ist es herauszufinden, wie die einzelnen Ensemblevorhersagesysteme, aus denen sich TIGGE zusammensetzt,
zur Vorhersagevariabilität in TIGGE beitragen. Es wird gezeigt, dass die
einzelnen Vorhersagesysteme einen unterschiedlichen Beitrag zur Vorhersagevariabilität und den daraus resultierenden Entwicklungsszenarien lie-
fern. Während manche der Vorhersagesysteme nur ein oder zwei mögliche Vorhersageszenarien erfassen und einen geringen Beitrag zur Vorhersagevariabilität leisten, zeigen andere Systeme beinahe alle Vorhersageszenarien, die in TIGGE zu finden sind. Insgesamt zeigt eine TIGGE
Multimodell-Ensemblevorhersage mehr mögliche Entwicklungsszenarien
als das EZMW- Ensemble. Diese volle Bandbreite an Entwicklungsszenarien in TIGGE wird allerdings nur erreicht, wenn das EZMW-Ensemble als
Mitglied in TIGGE mit berücksichtigt wird.

Ensemblevorhersagen zeigen mehrere mögliche Vorhersageszenarien für
die zukünftige Entwicklung einer bestimmten synoptischen Situation. In
Episoden mit verstärkter Vorhersageunsicherheit, wie z.B. während ET,
können diese Entwicklungsszenarien stark variieren. Verschiedene solcher

Vorhersageszenarien für ET-Fälle ermöglichen es, die physikalischen und dynamischen Prozesse während der außertropischen Umwandlung näher zu untersuchen. In dieser Arbeit wird ein Budget für die Eddy-Kinetische Energie in verschiedenen Vorhersageszenarien berechnet, die aus Vorhersagen des EZMW-Ensembles für zwei ET-Fälle extrahiert wurden. In einem Vergleich des Energie-Budget für die verschiedenen Szenarien zeigt sich der Beitrag des sich umwandelnden tropischen Wirbelsturms und der relevanten physikalischen Prozesse. Wichtige Ergebnisse sind, dass die Dauer der baroklinen Umwandlung der Eddy-Kinetischen Energie und die relative Phasenlage zwischen dem Sturm und der wellengeprägten Strömung in den mittleren Breiten die Auswirkung des Sturms auf die stromabwärts gelegene Strömung stark beeinflussen. Desweiteren scheint die Strömung der mittleren Breiten während eines kurzen Zeitintervalls besonders empfindlich auf die Wechselwirkung mit dem tropischen Wirbelsturm zu sein. Danach nimmt der Einfluss des oftmals weiterhin starken Sturms auf die Wellenstruktur in den mittleren Breiten ab.

In der vorliegenden Arbeit werden neue Aspekte der Vorhersagbarkeit und die physikalischen Prozesse während ET untersucht und mögliche Ansätze für weiterführende Studien aufgezeigt.

Contents

1 Introduction

Tropical cyclones rank among the most impressive and most hazardous weather systems on our planet. Each year a number of tropical cyclones develop above tropical and subtropical waters nearly all over the world. Fueled by ascending warm and moist air masses from the underlying oceans, tropical cyclones often develop into major cyclonic vortices with organised deep convection and hurricane force winds near their core region during their typical tropical life cycle. It is a well known fact that the landfall of a tropical cyclone threatens life and property. Thereby, the possible severe destruction in coastal areas is not only a consequence of the gale force winds themselves and the heavy precipitation, causing intense flooding, but also of the wind induced storm surges. But even after their tropical phase and a possible landfall, tropical cyclones still have great potential to cause high impact weather, particularly in regions which are typically not directly affected by a tropical cyclone.

Towards the end of their tropical life cycle, a considerable number of tropical cyclones undergo recurvature and head poleward into the mid-latitudes. Thus, the tropical cyclone abandons tropical regions and starts to interact with an environment, characterised e.g. by stronger wind-shear due to enhanced baroclinicity, and by lower sea surface temperatures. During its interaction with the mid-latitude flow, the tropical cyclone accelerates and experiences significant changes in its structural characteristics. Drier environmental air and increased vertical wind shear gradually disrupt the typical structure of the tropical system and initiate the transformation into an extratropical cyclone. This transformation process is referred to as extratropical transition (ET, Jones et al., 2003). If the transitioning tropical cyclone moves ahead of an upper-level trough or into another region that favours cyclogenesis, the ex-tropical storm may reintensify strongly into an extratropical cyclone with heavy precipitation and hurricane-force winds. Thereby, the phasing between the tropical cyclone and the mid-latitude wave pattern turned out to be crucial for the final outcome of the transitioning process. Sometimes the strongly reintensified ex-tropical cyclone subsequently traverses eastward through the whole ocean basin, reaching, for instance, Europe or western North America. In this case, the transitioned storm itself can cause high impact weather in these downstream regions.

It is not only the tropical cyclone itself, however, that experiences strong modification during ET. The interaction between the transitioning storm and the mid-latitude flow may cause significant changes in the extratropical flow pattern also, as the outflow of the transitioning storm at upper levels and the warm and moist tropical air masses can cause the amplification or even the development of an upper-level mid-latitude ridge directly downstream of the ET event. The amplification of the mid-latitude wave pattern in the vicinity of the transitioning storm may propagate further eastward. This causes the potential for strong cyclogenesis in regions far downstream of the ET event during the summer season, which is usually characterised by less cyclonic activity. Hence, an extratropical transition can also affect downstream regions like Europe or western North America indirectly.

The acceleration of the tropical cyclone during the ET, and the physical and dynamical processes at different horizontal scales involved in the interaction between the transitioning storm and the mid-latitude flow, often pose serious problems in numerical weather prediction for ET events. Hence, the extratropical transition and the associated potential for strong cyclogenesis in the downstream regions is accompanied by a decreased predictability for downstream regions, enforcing the possible impact of an ET event (Harr,

2010). Due to its often negative impact on predictability, the extratropical transition of tropical cyclones lies within the scope of the world weather research programme THORPEX (THe Observing system Research and Predictability EXperiment[1]), focusing on the improvement of 1-day to 2-week weather forecasts.

The enhancement of reliability in numerical weather forecasts during ET events requires a deeper understanding of the manifold underlying physical and dynamical processes and especially their impact on the predictability for downstream regions. Another possible approach to deal with the reduced predictability during ET or other high impact weather events is provided by developing new methods to better use and further improve existing forecast techniques, like ensemble forecasts. Such ensemble forecasts can provide useful information about the possibility of high impact weather and indicate possible synoptic developments, as these forecasts account for initialisation and partly also for model errors in numerical weather predictions. Furthermore, ensemble forecasts enable quantification of the inherent uncertainty during ET events. The present study deals with both of the aforementioned aspects on how predictability during ET can be increased.

In recent studies, Harr et al. (2008) and Anwender et al. (2008) investigated the predictability during ET in two operational ensemble prediction systems (EPS). They identified an increase in ensemble spread during the ET and extracted typical variability patterns among the individual ensemble members in the NCEP[2] and the ECMWF[3] EPS. A new multimodel ensemble forecasting system, the THORPEX Interactive Grand Global Ensemble (TIGGE) provides a means to extend the findings from Anwender et al. (2008) and Harr et al. (2008). In the first part of the study in hand, TIGGE is used to identify forecast uncertainty related to ET for the first

---

[1] http://www.wmo.int/pages/prog/arep/wwrp/new/thorpex_new.html
[2] National Centers of Environmental Prediction, USA
[3] European Centre for Medium-Range Weather Forecasts

time. Thereby, the focus is put on characterising the behaviour and the benefit of the multimodel ensemble approach in forecasting the ET of five tropical cyclones in the 2008 season.

An ensemble forecast provides a range of possible development scenarios for the ET of one specific storm. These different synoptic scenarios enable to examine the underlying physical processes during the ET causing the distinct developments. In the second part of this thesis, an analysis of the kinetic energy budget is conducted for clearly distinct development scenarios that resulted from an ECMWF ensemble forecast for the ET of two tropical cyclones. This approach enables to identify the role of the transitioning cyclone and the contribution from the extratropics during the modification of the mid-latitude wave pattern.

The thesis is structured as follows. Background information about extratropical transition as well as about ensemble forecasts is outlined in Chapter 2, followed by an introduction of the analysis techniques (Chapter 3). The core of the thesis present the investigation of predictability during ET events in TIGGE forecasts (Chapter 4) and an examination of the kinetic energy budget for selected development scenarios (Chapter 5). Conclusions are drawn in Chapter 6, completed by an outlook on open issues and possible future directions.

2 Theoretical Background

Different aspects of the ET of tropical cyclones are investigated in this study. This necessitates a brief introduction into the extratropical transitioning process which is provided in the first part of this chapter. The basic idea and the characteristics of ensemble forecasts, which form a powerful tool to examine various questions concerning ET, are introduced in the second part of this chapter. A short overview about the ET cases considered in this study closes the chapter.

2.1 Tropical Cyclones and their Extratropical Transition

Tropical cyclones can form over tropical and subtropical oceans in regions which fulfil distinct criteria, like high sea surface temperatures in the upper ocean layers (Palmén, 1948, Emanuel, 2003), weak vertical wind shear and other aspects (cf. Gray, 1968, Briegel and Frank, 1997, Tory and Frank, 2010). Surface heat fluxes emanating from the underlying ocean provide the dominant energy source for maintaining and further intensifying the tropical cyclone (Emanuel, 1986, 1991). During their whole life cycle, tropical cyclones are mainly steered by the environmental flow they are embedded in. Hence, in the Northern Atlantic and the Pacific basin there is a predominantly westward motion of the tropical cyclone along the southern flanks of the subtropical anticyclones. However, if the subtropical ridge weakens, an upper-level trough approaches or the tropical cyclone reaches the western boundary region of the ridge, the north- and northeastern wind component of the steering flow increases and might initiate the recurvature of the tropical cyclone.

2.1.1 Aspects of Extratropical Transition

If a tropical cyclone undergoes recurvature, with or without making landfall, it abandons the tropical regions favouring its maintenance and starts to interact with the mid-latitude flow pattern. The clearly distinct environmental conditions (Jones et al., 2003) in the extratropics exert a manifold influence on the tropical cyclone, leading to significant changes of its former tropical characteristics. During the interaction with the mid-latitude flow the tropical system undergoes ET and may transform into an extratropical cyclone. Furthermore, the tropical cyclone may start to interact with an extratropical synoptic disturbance (preexisting upper-level wave pattern and associated low-level baroclinic zone or extratropical cyclone). Under favourable conditions, the ex-tropical cyclone can completely transform and even reintensify into a strong extratropical cyclone. The other way round, the dominant anticyclonic upper-level outflow and the cyclonic rotation of the transitioning tropical cyclone itself can also markedly influence the mid-latitude wave pattern by amplifying, retarding or even exciting a Rossby wave train. As a consequence, this might precondition cyclogenesis in downstream regions, far away from the transitioning storm.

Extratropical transition is very common in nearly all ocean basins that favour the development of tropical cyclones. The processes involved in ET are assumed to be comparable in all regions. The number of transitioning tropical cyclones per year is closely related to the total number of tropical cyclones in the corresponding ocean basin. As highlighted by Jones et al. (2003), most ET events in the period between 1970 and 1999 took place in the western North Pacific, the region which has the largest number of tropical cyclones per year. On the other hand, most ET events in relation to the total number of tropical cyclones per year were found in the North Atlantic basin (46% of the North Atlantic tropical cyclones underwent ET, Hart and Evans, 2001). During their transition and potential reintensification as a strong extratropical cyclone with hurricane-force wind fields and heavy precipitation, the ex-tropical cyclones may traverse eastwards through the entire ocean basin (Thorncroft and Jones, 2000) and reach western Europe or the West Coast of North America. Some prominent and well-documented examples are the ET of Hurricane Lili in 1996 (e.g Browning et al., 1998, Agustí-Panareda et al., 2005), the ET of Hurricane Iris in 1995 (Thorncroft and Jones, 2000), or more recently the ET of Hurricane Hanna in 2008 (Grams, 2011), which affected the formation of a Mediterranean cyclone. The current state of knowledge about extratropical transition and challenges for further understanding of this complex process are summarised by Jones et al. (2003).

a) Structural Characteristics during Extratropical Transition

Based on the examination of satellite images for 30 ET events in the middle 1990s, Klein and co-authors developed a conceptual model for the ET of western North Pacific cyclones. This conceptual model divides the process into a transformation (Klein et al., 2000) and a reintensification (Klein et al., 2002) stage. The transformation process is partitioned further into three steps, during which the tropical cyclone reacts to the environmental changes and develops more and more extratropical characteristics. During the subsequent reintensification or extratropical stage (Jones et al., 2003), the ex-tropical cyclone can reintensify as an extratropical system or decay. Not all tropical cyclones that begin an extratropical transition complete the transformation stage, especially under conditions of strong shear or after landfall, and, in turn, not all transitioning systems enter the reintensification stage and develop into a (strong) extratropical cyclone.

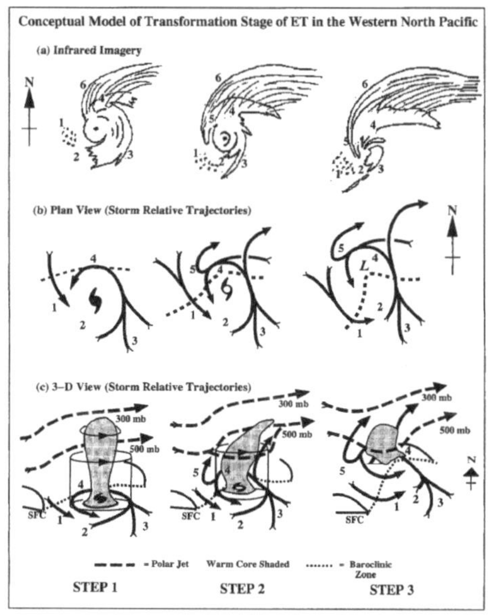

Figure 2.1: Conceptual model of Klein et al. (2000) for the three steps of the transformation stage during ET in the western North Pacific, based on examination of satellite images. a) schematic IR-imagery, b) schematic circulation pattern, c) schematic 3-d circulation. Numbers highlight different regions discussed in text. Figure 5 from Klein et al. (2000).

Figure 2.2: Satellite images for the ET of Super-Typhoon Ginger in 1997 indicating the structural changes during ET, as described in the conceptual model. a) 03 UTC 28 Sep, b) 00 UTC 29 September and c) 12 UTC 29 September 1997. Regions of interest are marked. White arrow indicates translation of the storm. Part of Figure 4 from Klein et al. (2000).

According to Klein et al. (2000), the tropical cyclone typically enters the transformation stage (step 1, Figure 2.1) when it starts to interact with the mid-latitude baroclinic zone and the associated vertical and horizontal wind shear. As a result of the interplay between the outer circulation of the tropical cyclone and the baroclinic zone, cold and dry environmental air from the mid-latitudes is advected cyclonically around the western flank of the transitioning storm (region 1, Figure 2.1). These dry air masses cause an erosion of deep convection and the associated rain bands in the western and southern outer quadrant of the transitioning storm (region 2). A dry slot forms in the western part of the southern quadrant and the appearance of the cloud and rain pattern in the satellite imagery becomes more asymmetric (Figure 2.2a). East of the circulation centre, warm and moist tropical air masses are advected into the northern and eastern quadrant of the cyclone (region 3), maintaining strong deep convection there (Figure 2.2a) and ascent along the baroclinic zone (region 4). These air masses interact with the adjacent polar jet, causing the formation of a northward extending cirrus shield (region 6 in the IR-scheme). At the same time, the transitioning storm typically accelerates due to the increased wind speed of the environmental flow.

If the ex-tropical cyclone is positioned directly to the south of the baroclinic zone it may enter step 2 of the transformation stage (Figure 2.1). The horizontal and vertical redistribution of air masses due to the cyclonic circulation continues to establish a dipole of lower tropospheric warm advection to the east (region 3) and cold advection to the west (region 1) of the transitioning storm. Warm (cold) advection to the east (west) further rotates the baroclinic zone into a more southwestern-northeastern expansion. Prevalent descending motion and the advection of cool, dry air masses on the western flank of the cyclone (region 1) further amplifies the development of the dry slot in the western quadrant and causes its expansion into the southern and already into the eastern quadrant (region 2). The cirrus cloud shield forms a sharp boundary region where the outflow encounters the polar jet (region 6). At the same time, parts of the ascending air masses east of the storm centre recirculate and descend in the northwestern part (region 5). Increased shear at the southern edge of the polar jet advects the upper-level part of the warm core downstream. This initiates the erosion of the tropical characteristics and the appearance of the transitioning system changes (Figure 2.2b).

The transformation stage is completed (step 3, Figure 2.1) if the ex-tropical cyclone becomes embedded in the baroclinic zone above regions with lower SST leading to a further weakening of deep convection in the transitioning storm. The continued dipole structure of the advection pattern (region 1 and 3), together with dry-adiabatically descending air masses, erode large parts of the eyewall (region 5), especially in the western quadrant of the former circulation centre. Advection of warm and moist air masses towards the baroclinic zone (region 4) triggers warm frontogenesis north and east of the transitioning storm (Harr and Elsberry, 2000). The expected cold frontogenesis in the southern and western quadrant is typically rather weak. In the end, the structure of cloud, temperature and humidity patterns of the transformed ex-tropical cyclone resembles the characteristics of an extratropical cyclone. This is also manifested in the typical asymmetric appearance in satellite images with a pronounced warm front and a possible "delta rain region" north of the cyclone (Figure 2.2c). The transformation into an extratropical cyclone also involves the development of a cold core structure in the centre of the storm, a feature that is not addressed in the conceptual model of Klein et al. (2000) as it can not be derived from satellite imagery.

After the transformation stage, the ex-tropical cyclone enters the extratropical stage if it did not decay during its transitioning stage. Under favourable mid-latitude conditions (i.e. an upper-level trough and its associated positive potential vorticity anomaly or a mature extratropical disturbance) the ex-tropical cyclone can (strongly) reintensify as an extratropical system.

A number of case studies (DiMego and Bosart, 1982, Harr and Elsberry, 2000, Harr et al., 2000) showed that the reintensification of the ex-tropical cyclone into an extratropical cyclone can be explained with the concept of a type-B cyclogenesis (Pettersen and Smebye, 1971). Therein, the interplay between positive vorticity advection in the upper levels and a frontal zone or disturbance, such as a transformed tropical cyclone, favours low-level cyclogenesis. Harr et al. (2000) further identified the role of the prevailing mid-latitude circulation pattern to be essential for the result and strength of reintensification. In one case, the transformed storm moves ahead of a mid-latitude trough that dominates the mid-latitude circulation (northwest-pattern). In the other case, the storm also moves ahead of a mid-latitude trough, but the mid-latitude flow pattern is dominated by another mature quasi-stationary cyclone northeast of the storm (northeast-pattern). The reintensification into a mature extratropical cyclone was found to be signif-icantly stronger, if the reintensification process proceeded in a northwest-pattern.

Another approach to define and classify the transitioning process was i.e. developed by Hart (2003) and Evans and Hart (2003). This method identi-fies the thermodynamic and asymmetric structure (warm- or cold core and vertical tilt) of the transitioning storm in analysis or forecast data sets and thus accounts for the development of the cold-core structure as well. A cold core and a vertical tilt are characteristics of typical extratropical cyclones. Hence, these parameters indicate the progression of the transformation pro-cess.

In addition to the dominant pattern of the mid-latitude circulation, the posi-tion of the transitioned storm relative to the mid-latitude wave pattern (re-ferred to as "phasing" between the ex-tropical cyclone and the mid-latitude wave pattern) and the amplification of the extratropical troughs and ridges also has a marked impact on the reintensification (e.g. Klein et al., 2002, Ritchie and Elsberry, 2003, 2007, Grams, 2011). The interaction between the storm and the mid-latitude flow is even complicated further as the strong outflow of the ex-tropical cyclone and the northward ascending warm and moist tropical air masses, transported around the eastern flank of the former circulation centre, can also modify the mid-latitude circulation pattern strongly. For example, the outflow can further amplify an upper-level ridge directly downstream of the transitioning storm, as it was shown in full (i.e. Grams, 2011) and idealised (i.e. Riemer et al., 2008) model simulations, or they can cause the formation of tropopause folds (Atallah and Bosart, 2003).

b) An Energetics Approach to Extratropical Transition

The first detailed examination of an extratropical transition was accomplished by Palmén (1958), who investigated the vertical circulation and the release of kinetic energy during the ET of Hurricane Hazel in 1954. In his study, the transitioning cyclone experienced enhanced generation of kinetic energy in direct thermal circulations with sinking cold and rising warm air masses. However, a large amount of this kinetic energy was exported away from the storm towards its surroundings, i.e. the mid-latitude flow, especially at upper levels. Subsequent studies analysed the kinetic energy budget for a number of other transitioning tropical cyclones (Kornegay and Vincent, 1976, Chien and Smith, 1977). They detected a detailed vertical distribution of gain and loss of kinetic energy due to cross-isohyptic or cross-contour flow (Kornegay and Vincent, 1976) from higher to lower heights by the ageostrophic wind and the associated work done by pressure forces, when air flows from high to low pressure, and the transformation of available potential into kinetic energy in direct thermal circulations. Furthermore, the strong export of kinetic energy in upper levels could be confirmed. DiMego and Bosart (1982) examined the kinetic energy budget for the extratropical transition of Hurricane Agnes (1972), whose reintensification closely resembled the type B cyclogenesis process (Pettersen and Smebye, 1971). In their study, DiMego and Bosart (1982) identified the cross-contour flow, mainly associated with the non-divergent wind, as a main source for kinetic energy in the transitioning storm, which apparently offsets its dissipation. Altogether, these studies highlight the potential of a kinetic energy approach to highlight the underlying physical and dynamical processes during extratropical transition, but they did not further examine these processes in detail.

During their investigation of the large-scale mid-latitude flow patterns and their influence on the reintensification process, Harr et al. (2000) also employed an analysis of the kinetic energy budget. The distinct mid-latitude flow features turned out to strongly influence the energy budget of the reintensifying ex-tropical cyclone. If the transformed storm undergoes reintensification in a northwest-pattern, the upper-level outflow towards the trough (acceleration towards lower pressure) results in a loss of kinetic energy in the vicinity of the transitioned storm. However, this loss is compensated and exceeded by baroclinic generation of kinetic energy during direct thermal solenoidal circulations, as the cyclone interacts with the mid-latitude trough, and by cross-contour flow mainly due to the non-divergent wind at upper levels within the cyclone. This net gain of kinetic energy supports the stronger reintensification of the cyclone in a northwest flow pattern. If the transitioned storm moves into a northeast-pattern, the non-divergent wind, whose cross-contour flow is a primary source of kinetic energy in the northwest-pattern, will be directed towards the dominant quasi-geostrophic low pressure system in the eastern part of the ocean basin. This constitutes sink of kinetic energy in the reintensifying storm and apparently hinders the reintensification of the ex-tropical cyclone into a strong extratropical cyclone.

Some years later, Harr and Dea (2009) revisited the kinetic energy budgets during ET events. They adapted the idea of "downstream baroclinic development", formulated by Isidoro Orlanski and co-authors in the early 1990 (Orlanski and Sheldon, 1995), to investigate the ET of five tropical cyclones in terms of their kinetic energy distribution with respect to a mean (climatological) distribution of kinetic energy (this budget is referred to as eddy kinetic energy budget, where eddy states for the investigation of deviations from a mean state). During a closer examination of the individual terms influencing the eddy kinetic energy budget, Harr and Dea (2009) were able to identify different interactions between the transitioning and reintensifying ex-tropical cyclones and the mid-latitude flow. At the same time the eddy kinetic energy budget also indicates the influence of the tropical cyclone on the mid-latitude flow pattern. Strong fluxes of eddy kinetic energy are identified to emanate from the ex-tropical cyclone and to amplify the mid-latitude wave pattern. According to the idea of "downstream baroclinic development", the ex-tropical cyclone acts as an additional source of kinetic energy for the further amplification of the wave pattern even in regions that typically have low baroclinicity and hence less potential for the development and amplification of a mid-latitude wave pattern. A more detailed introduction into the basic idea of "downstream baroclinic development" is provided in Section 3.2.

c) Predictability of Extratropical Transition

The preceeding sections pointed to the various physical and dynamical processes influencing the transition and reintensification of a tropical cyclone into an extratropical cyclonic system. Especially the sensitivity of reintensification to the phasing between the transitioning storm and the mid-latitude wave pattern poses serious problems in the accurate prediction of ET events using numerical forecast models. The correct representation of the structural changes of the ex-tropical cyclone during its ET can also raise serious problems in numerical forecasts during ET (Evans et al., 2006). Hence, the

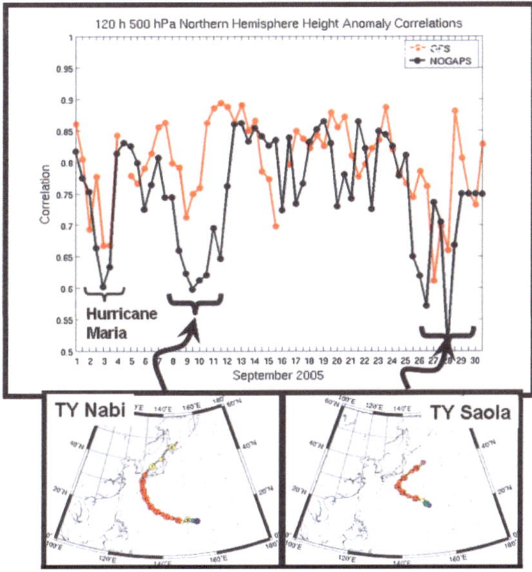

Figure 2.3: Anomaly correlation coefficient for 5 day forecast of 500 hPa geopotential height over the Northern Hemisphere during the ET of three tropical cyclones (Maria, Nabi and Saloa) in 2005. The ET of the tropical cyclones caused a significant drop in predictability. Red: Global Forecast System from the National Centres for Environmental Prediction (NCEP-GFS). Black: Navy Operational Global Atmospheric Prediction System (NOGAPS). Figure 1 from Harr (2010).

accuracy of numerical forecasts is often reduced during ET events, as it is depicted by i.e. Harr (2010) (adapted here as Figure 2.3). The forecast skill of the global numerical models (NCEP GFS[1] and NOGAPS[2]), expressed as the Anomaly Correlation Coefficient (ACC, cf. Section 5.3.1) for the Northern Hemisphere is clearly decreased during the ET of three tropical cyclones in September 2005. Hence, the extratropical transition of a hurricane in the North Atlantic basin can cause a significant drop in forecast accuracy for the Atlantic-European Sector.

---

[1] National Center for Environmental Prediction (USA) Global Forecast System

[2] Navy Operational Global Atmospheric Prediction System

The physical and dynamical processes within the tropical cyclone and its surroundings (boundary layer, air-sea interaction, friction) during ET are yet not well understood. Many of these processes that determine the structure of the tropical cyclone and its outflow during the ET, such as moist processes like deep convection and the associated latent heat release, evaporation, turbulent transfer of moisture, or processes like gravity waves and friction act on a subgrid scale for most numerical forecast models. Hence, these processes have to be parametrised in the model formulation, which means that the contribution of the these subgrid scale is deduced from variables at scales that can be resolved by the model. Without detailed knowledge about the processes and their respective contribution during ET, the parametrisation of these processes in numerical models states a source for additional forecast errors.

Another aspect limiting predictability during ET is the interaction of different horizontal scales (Jones et al., 2003). While the mid-latitude flow and the embedded extratropical cyclones span a sufficient area (5000 km diameter) to be correctly resolved and captured by global numerical forecast models, the rather small scale tropical cyclones (500-1000 km diameter) and especially their intensity and track are often misrepresented in models, whose spatial resolutions typically ranges in the magnitude of 50 km. Limited area models, which would allow to resolve the tropical cyclone and its interaction with the mid-latitude flow to a better extent, require correct initial and boundary conditions (cf. Section 2.2) to resemble the mid-latitude flow pattern the cyclone approaches. However, if the tropical cyclone is misrepresented in the global model, the initial and boundary conditions will introduce uncertainties in the limited area models, limiting their skill during ET as well.

A correct representation of the phasing between the mid-latitude flow and the approaching tropical cyclone in the initial conditions is a crucial factor for accurately predicting the future atmospheric state during ET. In some experiments, Grams (2011) could show that a shift of approximately 100 km in the position of Typhoon Jangmi relative to the mid-latitude trough determined the reintensification or decay of the ex-tropical cyclone during ET. However, the general acceleration of the transitioning cyclone during its interaction with the mid-latitude flow and the onset of the acceleration is hard to predict and even hard to analyse. Thus the correct identification of the current cyclone position poses a serious problem during the initialisation of the forecast. For example, if the tropical cyclone accelerates strongly just after the analysis, this acceleration might not be captured in the numerical model and the interaction between the transitioning cyclone and the mid-latitude flow might be misrepresented. This might cause completely different synoptic developments. As the transitioning tropical cyclone itself is capable of strongly influencing the mid-latitude flow by modifying the trough-ridge pattern, the decrease in predictability during ET events can also effect regions far downstream (Jones et al., 2003, Anwender et al., 2008, Harr et al., 2008, Harr, 2010).

2.2 Numerical Weather Prediction and Ensemble Forecasts

The first attempt to mathematically predict the future state of the atmosphere by solving a differential equation, was undertaken in 1922 by Lewis Fry Richardson. Since then, numerical weather forecast became a large topic in atmospheric sciences and underwent a strong increase in accuracy during the past 30 years (Simmons and Hollingsworth, 2002). Large parts of this increase are attributable to the rapid development in computational capacity. Nevertheless, time and time again atmospheric situations occur, like i.e. an ET event, during which the numerical weather predictions fail. Furthermore, the atmospheric state itself is predictable just to a certain extent, as we will see later.

The subsequent sections first give a short overview about numerical weather predictions and its natural limitations. Thereafter, a different approach in numerical weather predictions, so-called ensemble forecasts, are introduced and described in more detail, as ensemble forecasts form the data base for the study in hand.

2.2.1 Basics of Numerical Weather Prediction

The actual state of the atmosphere at each individual point at a specific time can be expressed using a state vector, which contains the current value of an adequate set of hydro- and thermodynamic state variables like temperature, surface pressure, wind velocity or specific humidity, and describes the phase space of the system. The temporal evolution of the atmosphere can then be described by a sequence of state vectors between the initial and the future state, which form the trajectory of the system through the phase space (Lorenz, 1963). With knowing a set of physical equations, the initial state of the atmosphere and the conditions on its boundary regions, the future evolution of the atmospheric state is predictable. These physical equations describe the conservation of momentum, mass, energy and water mass, and furthermore involve the equation of state for ideal gases (Kalnay, 2003).

To predict the future state of the atmosphere numerically, the underlying differential physical equations must be expressed in terms of the state variables and approximated by difference equations utilising a suitable discretisation. Furthermore, physical processes like radiation, turbulence and convection, must be parametrised for numerical weather forecasts. These processes take place on smaller scales as to which the governing equations can be discretised due to limited computational resources and time in an operationally running forecast system. However, they are very important for the development at the larger scales, which are captured by the forecast model. A correct parametrisation of these processes is very important for a successful forecast, but it is also one of the most sophisticated parts of numerical weather prediction (Buizza, 2003).

The actual state of the atmosphere can be determined by the use of a variety of observations, i.e. weather stations at land and on ships or satellite measurements, to name but a few. These observations are spread inhomogeneously over the whole globe and are conducted on different vertical levels. Satellite observations are characterised by nearly global coverage, but they mainly provide information about specific parts of the atmosphere and can often be conducted only in cloud-free regions, while weather stations may provide multi-level data from upper air sounding, but only at specific and inhomogeneous distributed points, preferably over land. Furthermore, the observations are taken at specific synoptic hours or even continuously. To predict the future state of the atmosphere, these observed variables must be transformed into the atmospheric state variables of the model and interpolated on discrete and homogeneously distributed grid points to be included in the forecast process. This is done by a process, which is referred to as data assimilation.

Each initialisation of a numerical forecast model is preceded by a data assimilation cycle. During the data assimilation procedure, a short range forecast of the numerical model is modified to be as close as possible to the observations by minimising a cost function under consideration of the background and observational error covariances. Thereby, observations are incorporated that are taken during a predefined time slot around the initialisation time. This modified short range forecast then forms the best possible approximation to the actual state of the atmosphere, which provides the initialisation state for the numerical forecast model (cf. Kalnay, 2003, Buizza, 2003, Bouttier and Courtier, 1999, Daley, 1991). However, due to restrictions in observations and model solutions, it will probably never be possible to know the absolute "truth" of the atmospheric state.

Basically, there exist two main error sources in numerical weather prediction systems. Uncertainties in the initial conditions are the first error source limiting the forecast skill. These uncertainties mainly arise from the improper observation of the actual atmospheric state because of systematic and random instrument errors, the inhomogeneity and low resolution of the global observing network and the approximations in assimilating the data into the model (cf. Shapiro and Thorpe, 2004). During the forecast these initialisation errors grow through the dynamics of the model and may account for a large part of the forecast error on the shorter forecast range (Leutbecher and Palmer, 2008). In addition to these so-called initial condition errors, a numerical weather prediction also suffers from the attempt to describe the atmospheric flow itself in terms of a numerical model. Errors are also introduced by the discretisation of the underlying equations and the assumptions made in parametrisation of small- scale processes. Those errors are referred to as model errors.

The description above suggests that if we knew the actual state of the atmosphere perfectly and had a forecast model without any errors, we would be able to make a perfect forecast, following the principle of determinism. However, in 1963 Edward N. Lorenz identified the chaotic nature of the atmospheric flow (Lorenz, 1963, 1962). During model simulations he discovered that simply rounding down the state variables during initialisation of the model caused two forecasts to develop more and more distinctly after some forecast hours. Finally, the two forecasts described two totally different atmospheric states. His investigations led him to the assumption that there exists a natural limit in predicting the future state of the atmosphere, which he estimated to be around two weeks. Furthermore, he supposed this natural limit to be dependent on the stability of the underlying flow. If the flow regime is unstable, infinitesimal errors in the initial conditions, such as simple round-off errors, will strongly affect the outcome of the forecast and the predictability would vary from time to time and region to region

(Lorenz, 1982). These impressions induced Lorenz to suggest an alternative statistical approach to numerical weather prediction, as it was already presumed some years before by E. Eady (Eady, 1951).

One way to respond to these limitations of a deterministic weather forecast is provided by combining the dynamical principles with a stochastic approach. In a stochastic sense, the present state of the atmosphere only determines a probability distribution for the future states of the atmosphere (Lorenz, 1982), and the deterministic forecast is just one realisation for the future state, beside many others. This basic concept led to the development of ensemble forecasts, which have been running operationally since the early 1990's.

2.2.2 Ensemble Forecasts

The basic idea of ensemble forecasting is to assess the probability of occurrence for future states of the atmosphere by carrying out several parallel runs of one forecast model with slightly modified initial conditions or model perturbations. Each run provides one possible solution for the future state. The differences between these several future states indicate, how predictable the situation is (large differences coincides with poor predictability) and how probable the different solutions are to occur.

a) Basic theory

After the early ideas from Eady (1951) and Lorenz (1963), Edward Epstein (Epstein, 1969) was the first to develop a concrete idea for how to set up an ensemble prediction system, which he referred to as the stochastic-dynamic approach (cf. Lewis, 2005). His basic ideas and findings provided the base for the future ensemble prediction systems under restricted computational resources.

Generally, in the ensemble approach to numerical weather prediction, the uncertainty in the initial state of the atmospheric system is expressed in terms of a probability density function (pdf). The temporal evolution of this pdf under the forecast model in the phase space of the system can be described by Liouville's equation, a continuity equation for probability. Basically, this equation is closely related to the continuity equation of mass, i.e. stating that the total amount of probability remains constant, which is equivalent to the fact that no ensemble member can be created or destroyed (Epstein, 1969, Kalnay, 2003). A direct numerical integration of this equation at incremental values appeared to be impossible for a system with as many degrees of freedom as the atmosphere, because the actual value of the state variables at each grid-point must be stored and updated each time step.

Current operational ensemble prediction systems are based on a Monte Carlo technique which was adapted for the application in numerical weather prediction by Leith (1974). In a conventional Monte Carlo method, the initial pdf is gained through a random sampling of the initial state and is known precisely. Thereafter, this certain initial pdf is non-linearly evolved in time, using a forecast model. The evolved pdf then describes the probability distribution of the system at the final time of integration, expressed in the phase space of the system. However, in the case of weather prediction this conventional approach does not work. On the one hand, the initial pdf will never be defined precisely, as the errors that determine this initial pdf are only known approximately (Buizza, 2003, Leutbecher and Palmer, 2008). On the other hand, with the current computer resources it is impossible to integrate the fully determined initial pdf for a system as the atmosphere[3] (Leutbecher and Palmer, 2008). This problem is overcome by only sampling parts of the initial pdf and developing this finite number of sample members, referred to as ensemble members, into the future.

---

[3] a pdf for each variable at each grid-point and level would require a numerical integration to fully determine the probability distribution of the system in the future

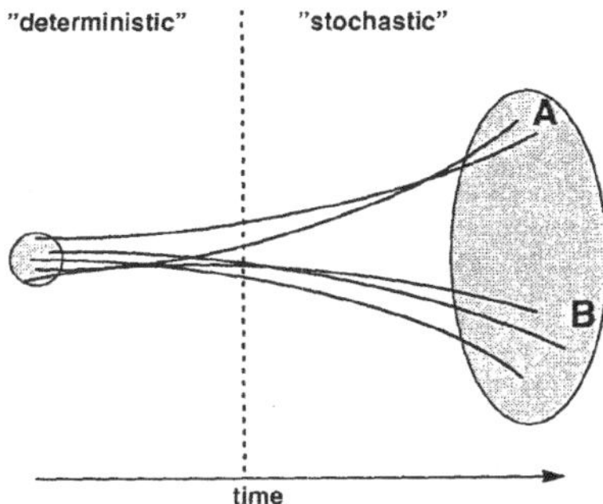

Figure 2.4: Example for an operational ensemble forecast. A representative set of perturbed initial conditions with some uncertainty distribution (circle) is advanced forward in time, indicated by the trajectories. Towards the end of the forecast the ensemble members span a range of possible solutions, represented by the ellipse. The more members point to one solution, the higher is the probability for this solution to occur. At the beginning, while the trajectories lie close to each other, the development can be considered as deterministic, thereafter it is stochastic. From Tracton and Kalnay (1993).

The state of all sample members at the final time then forms an approximation of the pdf at the later forecast time (Figure 2.4). Furthermore, the distribution of the pdf in phase space provides information about the probability of the scenarios to occur (Tracton and Kalnay, 1993). In this way only a part of the uncertainties in all degrees of freedom of the atmospheric system is considered. Hence, due to computational restrictions, the chosen ensemble members must be constructed to cover a sufficient part of the future pdf with the least possible number of members, to turn the ensemble forecast into a beneficial tool for forecasters to predict the future state of the atmosphere. In simple terms this means that the ensemble forecast should furnish a range of possible solutions for the future atmospheric state, which contains the real development (verifying analysis), but that does not span the whole range of climatologically feasible atmospheric states. This would imply that every state is possible, but would not provide any additional information. In contrast, if the range of possible solutions does not contain the verifying analysis, or if it is too narrow to describe the real variability of the atmospheric state, the ensemble forecast is under-dispersive and may mislead the forecaster to wrong assumptions about the future state of the atmosphere. The uncertainty of the forecast (i.e. the range of possible solutions), the so-called ensemble spread, acts as a measure for forecast uncertainty. The ensemble approach is confirmed to be valuable by the finding that a conventional deterministic forecast often is less skillful than a forecast which is provided by the mean of an ensemble of forecasts (Kalnay, 2003).

One of the essential components for a well-performing EPS fulfilling the criteria stated above are the initial perturbations. They should resemble the analysis errors that arise during the determination of the initial state for the forecast and should cover the strongest variability in the atmospheric system, i.e. those parts of the atmospheric flow that allow strongest error growth and thus have most impact on the future state of the atmospheric system (Kalnay, 2003). Another important point for the performance of the EPS is whether and by which method model errors are taken into account during the forecasting process. The consideration of these model errors should resemble uncertainties that arise due to approximations in the parametrisation schemes for subgrid scale processes or due to uncertainties in the model formulation itself. Finally, the ensemble forecast is only useful, if it is available in a certain time frame. It is clear that a two-week forecast would be useless if its computations requires 14 days. In the case of ensemble forecasts, the computation time is certainly governed by the number of members contained in the ensemble. The typical number of ensemble members in operational models ranges from about ten to 50. But the more members are included, the more reliable gets the information and the spread of the ensemble system. To be able to run a sufficient number of members, the model setup, creation of initial perturbation, parallelisation etc. must be as efficient as possible. However, although research during the recent years led to marked technical advances in these fields, the current operational global ensemble forecasts still have coarser resolution than the deterministic global forecasts and thus tend to neglect processes on smaller scales.

b) Perturbation Methods

During the recent 20 years an increasing number of operational weather services around the world started to include an ensemble forecasting system in their repertoire of numerical weather prediction systems. The challenges in creating an ensemble that fulfils the above mentioned criteria led to an advanced research in the field of ensemble forecasting. Several techniques for generating the initial perturbations and considering the model errors were established. A brief overview about the most common methods is provided here.

Initial Perturbations

Generally, initial perturbations are constructed to affect those parts of the actual atmospheric flow that will lead to the strongest error growth, in order to be able to cover a large fraction of the uncertainty in the forecast. These structures are then perturbed during the initialisation of the forecast to obtain the best possible approximation of the probability density function of the future atmospheric state.

Singular Vectors

One possibility to identify unstable regions in the atmospheric flow is provided by the use of atmospheric singular vectors. The singular vector formalism aims to identify those regions of the flow field that allow strongest error growth in a specified time window (the so-called optimisation time), e.g. 48 hours, in a linear framework. The perturbation growth is thereby measured with respect to a predefined norm, e.g. the total energy of the atmospheric flow. Singular vectors can also be determined for predefined regions of the atmospheric flow, i.e. for the surroundings of a tropical cyclone. To generate perturbations to the initial conditions, the amplitude of the singular vectors is scaled to resemble the magnitude of estimates of the analysis uncertainty. The singular vector method leads to adequate initial perturbations that allow reasonable error growth during the forecast period and cause the spread among the several members often to be sufficient. The method was shown to perform slightly better than the so-called breeding method in a direct comparison for one ensemble prediction system (Magnusson et al., 2008). However, the generation of singular vector initial perturbations requires costly computations and necessitates a linearised version of the forecast model and its adjoint. More details about generating initial perturbations with the singular vector method are provided by Buizza and Palmer (1995), Kalnay (2003), Buizza (2006), Leutbecher and Palmer (2008) or, briefly, in Verret (2010).

Breeding

The breeding method identifies growing perturbations by subsequent forecast cycles from a randomly perturbed initial state. It starts with adding a random initial perturbation to a forecast run or a sequence of analyses to create a perturbed initial condition. Thereby, the perturbation is scaled to represent a norm that is typical for analysis error (i.e. root mean square error of the geopotential height field). Now two forecast runs of the full nonlinear forecast model are performed, whereas one starts from the unperturbed analysis and the other one from the perturbed initial condition. After a given time interval, e.g. 6 hours, the difference between the perturbed and unperturbed run is determined, scaled back to the magnitude of the former initial condition and then employed as a new initial condition for a subsequent forecast run. Approximately 3-4 days after the breeding cycle was initialised, the perturbations in the breeding cycle typically grow faster than just randomly added perturbations during the subsequent forecast intervals and are apparently independent from the scaling measure (rms) and scaling period (Toth and Kalnay, 1993, 1997). Strongest growing initial perturbations then indicate the most unstable regions of current atmospheric flow. In the operational application of the breeding method the initial perturbations for a new ensemble run are extracted directly from the ensemble forecast itself. The scaling factor of the perturbation can be adapted to allow larger error growth rates over sea than over land, to meet the increased uncertainty in these regions, as a consequence of the lack of observations. All in all, the breeding method also leads to reasonable initial perturbations and has the benefit of not requiring any additional computations, making this method competitive with respect to computational costs. More details are provided e.g. by Tracton and Kalnay (1993), Toth and Kalnay (1997) and Kalnay (2003).

Ensemble Kalman Filter and Ensemble Transform Kalman Filter

Another approach to create initial perturbations is provided by the use of ensemble data assimilation methods. In general, an analysis can be created by updating a first-guess or background atmospheric state (short-range forecast with the full model from an earlier analysis) by a set of observations. Thereby, the observations are statistically interpolated onto the background field, under the consideration of the error covariances for the background field and for the observation fields. For a linear system, the background covariance matrix can be gained by a Kalman Filter (Kalman, 1960). For complex nonlinear systems like the atmosphere, the background error covariances can be approximated by using an ensemble of background forecasts, which is referred to as Ensemble Kalman Filter (EnKF). This ensemble of background forecasts is based on an ensemble of analyses, which are evolved in time with different versions of the forecast model to account for model errors. Now there exist different ways to create the new ensemble of analyses, which then can be used as initial conditions for an ensemble prediction system. One obvious possibility is provided by adding random perturbations to the observations, before they are assimilated and combined with the background field during the data assimilation cycle. However, this necessitates a rather large ensemble set to get an adequate approximation of the background error covariance. Another way is provided by the use of Kalman square-root filters, like the Ensemble Transform Kalman Filter (ETKF). Thereby, observations are only assimilated into the ensemble mean to provide a mean analysis. The perturbations to create the analysis ensemble are then achieved by transforming the perturbations from the background (or former forecast) ensemble into analysis perturbations. Since only the ensemble mean is updated, the ETKF approach is computationally inexpensive, but its mean analysis often is less accurate than the mean of the full EnKF approach. An introduction into the ETKF method is for example provided by Kalnay (2003), Ott et al. (2004), Hamill (2006), Kalnay (2009) and Verret (2010). More details about the EnKF technique can be found in e.g. Houtekamer et al. (1996), Houtekamer and Mitchell (1998) and Houtekamer and Mitchell (2005), while the ETKF technique is e.g. described by Bishop et al. (2001) and Wang and Bishop (2003).

Empirical Orthogonal Function (EOF)-Based Perturbations

Initial perturbations can also be gained from an EOF analysis of a set of short-range forecasts. Random perturbations are added to a control analysis, to get a first set of perturbed initial states. The different initial states and the unperturbed analysis are then forward integrated for a short-range time interval (e.g. 36 h), using the full numerical model. Thereafter, the control forecast is subtracted from the different perturbations at defined time steps, which results in a time series for the forecast-difference fields. By executing an EOF analysis for these difference fields, the leading eigenvectors indicating the strongest growth can be extracted. These eigenvectors are then added to or subtracted from the control analysis to create the initial ensemble perturbation for the full ensemble forecast. See Zhang and Krishnamurti (1999) for more information.

Perturbation Methods to Account for Model Errors

Model errors can be addressed by the use of so-called "stochastic physics". In one such method the parametrised tendencies for the model variables are disturbed by adding random perturbations, whose magnitudes are proportional to the unperturbed tendencies (Shutts, 2010). This results in spatially and temporarily correlated random perturbation patterns (see e.g. Palmer, 2001, Palmer and Williams, 2008). More recently, additional stochastic physic schemes have been developed to consider the loss of kinetic energy during the forecast cycle. Basically, kinetic energy is dispersed to subgrid scales when circulation patterns decay to smaller scales. The kinetic energy backscatter scheme re-introduces this kinetic energy into the flow to prevent unnatural damping of the flow structure (Shutts, 2005, Berner et al., 2009).

c) Multimodel Ensembles

Even if the model errors are addressed adequately by using an ensemble analysis scheme or introducing stochastic perturbation to the physics of the model and even if the initial perturbations span a sufficient range of possibilities, there might still exist some error sources that are not addressed in an individual EPS and that may cause this ensemble forecast to fail in predicting the real development. These errors can, for example, result from errors in the model formulation which are still not considered (e.g. due to discretisation or approximations in parametrisation schemes), from uncertainties during generation of the initial perturbations or rather obviously from the coarse horizontal and vertical resolution. This led to recent investigations concerning the creation of so-called multimodel ensembles, in which different individual deterministic models or even ensemble prediction system are brought together to form a new combined ensemble forecast system. This allows to further consider uncertainties in the forecast that arise due to the model formulation or the generation of the initial perturbations. If, for example, one of the individual ensemble forecasts is under-dispersive in a specific case, this might be compensated by one of the other ensemble forecasts. Then the multimodel ensemble will be able to predict this future event, such as a case of high-impact weather, while the single ensemble system would fail.

The THORPEX Interactive Grand Global Ensemble (TIGGE)

The most extensive approach to create a multimodel ensemble in the previous years was undertaken in the framework of THORPEX: Ten global ensemble prediction systems, running at meteorological centres around the world, are brought together to form the THORPEX Interactive Grand Global Ensemble (TIGGE) (Bougeault et al., 2010). This multimodel ensemble was initiated in 2005 and became fully operational in 2008.

Each of the ten contributing meteorological centres delivers its common operational ensemble forecast and the corresponding control forecast in near real time to a data base hosted by three "archive centres". During a postprocessing at these "archive centres", the individual ensemble forecasts are converted to a common grid and data format and are interpolated onto the same pressure levels. After approximately two days, these individual but adjusted ensemble forecasts are then freely available for research and educational purposes. The individual ensemble forecasts can be arbitrarily combined to form a multimodel ensemble. This multimodel ensemble then provides the base for manifold investigations, from single case studies to predictability investigations for the whole sample period.

Table 2.1: Configuration and overview of the ten EPS combined in the TIGGE data set. Based on the information provided by the respective meteorological centres in the ECMWF TIGGE archive (http://www.tigge.ecmwf.int). L refers to the number of vertical levels, T stands for spectral triangular truncation at the indicated wave number and TL for spectral triangular truncation at the indicated wave number but with linear grid

Centre	Members	Initial. Time (UTC)	Forecast Days	Initial Pert.	Model error	Forecast Resolution
Australia (BOM[1])	33	00, 12	10	SV	No	TL119 L19
Brazil (CPTEC[2])	15	00, 12	15	EOF-based	No	TL126 L28
Canada (CMC[3])	21	00, 12	16	EnKF	Stoch. Phys.	0.9deg L28
China (CMA[4])	15	00, 12	10	SV	No	T106 L19
Europe (ECMWF[5])	51	00, 12	15	SV	Stoch. Phys.	T399 L62(0-7d)
France (MeteoFrance)	11	18	2.5	SV	No	TL358 L41
Japan (JMA[6])	51	12	9	SV	No	TL319 L60
Korea (KMA[7])	17	00,12	10.5	BV	No	T213 L40
UK (UKMO[8])	24	00,12	15	ETKF	Stoch. Phys.	0.83deg lat × 1.25deg lon L38
USA (NCEP[9])	21	00,06,12,18	16	ETR	No	T126 L28

[1] Bureau of Meteorology
[2] Centro de Previsao de Tempo e Estudos Climaticos
[3] Canadian Meteorological Centre
[4] China Meteorological Administration
[5] European Centre for Medium-Range Weather Forecasts
[6] Japan Meteorological Agency
[7] Korea Meteorological Administration
[8] UK Meteorological Office
[9] National Centers for Environmental Prediction

The only restriction in combining the different individual ensembles arises from the initialisation times of the forecasts (some models are initialised every 6 hours, while others are run just once a day) and from the availability of some forecast parameters (e.g. potential temperature on a potential vorticity level is only provided by three of the ten centres). Standard parameters like geopotential height, surface pressure, temperature etc. are available from all ensemble forecast systems. Such a wealth of data from operational numerical weather prediction systems provides new opportunities to advance our knowledge of the predictability of specific weather systems and has not previously been available in the research community. A short overview of the ten ensemble systems and their individual configurations in 2008 is provided in Table 2.1. More details on the individual ensemble configurations and continuative references are e.g. provided in the TIGGE archive, hosted by the ECMWF (http://www.tigge.ecmwf.int).

2.3 Ensemble Forecasts for Extratropical Transition

Due to the diversity of the uncertainties that may hamper a reliable prediction of the ET event (Section 2.1.1), ensemble forecasts are particularly suited for predicting ET. Ensemble forecasts constitute an overview about the possible synoptic developments that can arise from the actual atmospheric state. If these scenarios are closely related, the occurrence of this scenario is likely. On the other hand, if the scenarios vary conspicuously, the future development of the atmosphere is hard to predict and depends strongly on a correct formulation of the initial and boundary conditions. In addition to this probabilistic information about the future state of the atmosphere, ensemble forecasts also allow the factors influencing predictability to be studied. As they provide a number of solutions for the future development of the actual synoptic situation the physical and dynamical processes that cause the distinct development scenarios to be different can be examined, at least to a certain degree.

Using global ensemble forecasts from NCEP and ECMWF for a number of ET events, Harr et al. (2008) and Anwender et al. (2008) could show that the decrease in predictability during ET is associated with an increase in variability among the ensemble members. Concerning the reason for the loss in predictability, they identified the uncertainty about the amplification of the mid-latitude wave pattern during the ET of Typhoon Nabi (2005) to cause increased variability among the ensemble members in downstream regions (Harr et al., 2008). Thereby, the predictability decreased after the completion of ET. If the forecast initialisation time approached the actual ET time, the variability among the ensemble members decreased and thus the predictability for downstream regions increased. Although the investigated storms showed different characteristics during their interaction with the mid-latitude flow, Anwender et al. (2008) could identify dominant and recurring variability patterns. Those regions of increased variability were related to specific features of the mid-latitude flow during the ET of the investigated tropical cyclones. The strongest uncertainty in the ensemble forecasts was tied to the amplitudes of the troughs and ridges (amplitude pattern) and the position of the trough and ridge flanks (shift pattern). An amplitude pattern is caused by a strongly varying representation of the amplification of the wave pattern in the individual ensemble members, i.e. a strongly developed trough in one members and a ridge in another. In the case of the shift pattern the strongest differences among the members are located at the flanks of the wave pattern, indicating a more eastward position in one and a more westward position in another ensemble member. In general, some mixtures of both patterns can be found as well. This clearly confirms the ability of the transitioning tropical cyclone to influence the amplification and propagation (i.e. by deceleration) of the extratropical wave pattern, respectively.

2.4 Selected Extratropical Transition Events

The extratropical transitions from six different tropical cyclones are examined in the present work. The first part of this study considers the predictability during ET events in the new TIGGE multimodel ensemble (Chapter 4). Therefore, ten forecasts for the ET of five different tropical cyclones are investigated to characterise the behaviour of the new data set during ET. For this part of the study the selected forecasts are initialised prior to the ET[4] of the considered tropical cyclones and show increased forecast uncertainty after the ET in each case. In the second part of the study, the ET of two tropical cyclones is examined in more detail using an analysis of the eddy kinetic energy budget (Chapter 5). The six ET cases considered in the study are introduced briefly in the following section.

2.4.1 Hurricane Hanna

Hurricane Hanna (Figure 2.5) formed from a tropical wave in the eastern Atlantic Ocean around 19 August 2008. During the subsequent days the convective system intensified and was declared as tropical storm at 12 UTC 28 August 2008. The system approached a low-shear environment that resulted from a decaying upper-level low and underwent strong intensification. It reached hurricane strength around 18 UTC 1 September 2008 as it was located near the Caicos Islands. Shortly after the strong reintensification, Hanna moved into a region with enhanced shear in the vicinity of a ridge over the US and started to weaken. During the subsequent days, Tropical Storm Hanna interacted with an upper level low and made a counterclockwise loop near Caicos and Haiti, where it caused 500 fatalities. The system then approached the periphery of the subtropical ridge and accelerated in a northward direction. It made landfall near the border between

---

[4]declaration of the system as extratropical cyclone by the Regional Specialised Meteorological Centre (RSMC)

North and South Carolina around 07 UTC 6 September 2008 as tropical storm, causing flooding and power disruptions. Thereafter the weakening system moved north-northeastwards and merged with a cold front over southern New England. Hanna was declared as an extratropical system (Ex-Hanna) on 12 UTC 7 September 2008. As the system moved offshore east of Newfoundland Ex-Hanna further weakened as it interacted with another frontal boundary. Later on 8 September 2008 the remnants of Ex-Hanna were located ahead of an approaching shortwave trough and start to reintensify as an extratropical cyclone around 12 UTC 09 September 2008. This system then propagated towards Europe and supported the formation of a cut-off cyclone in the Mediterranean (Grams et al., 2011). Detailed information about Hurricane Hanna is provided in the Tropical Cyclone Report by Brown and Kimberlain (2008). An ensemble forecast for Hurricane Hanna is examined in the TIGGE study (Chapter 4). The ET of Hurricane Hanna is also examined in the framework of the eddy kinetic energy analysis (Section 5.3.1).

2.4.2 Hurricane Ike

Hurricane Ike (Figure 2.6) emanated from a tropical wave near the West Coast of Africa around 30 August 2008. It strengthened to tropical storm status until 12 UTC 1 September 2008 near the Cape Verde Islands. The subtropical ridge played a crucial role in the steering of Hurricane Ike's track. Initially, the storm moved west-northwestwards along the southern edge of the subtropical North Atlantic anticyclone. Rather dry environmental air masses suppressed deep convection and may have hindered a strong intensification of Ike. As the ridge weakened slightly the storm intensified and reached category 4 at 6 UTC 4 September 2008. The restrengthening subtropical ridge in the western North Atlantic forced Ike into a more west-southwestward motion, and caused a slight weakening of the hurricane. Due to its southwestward motion Ike then approached a region with low shear, reintensified as category 4 hurricane and moved towards Cuba. Ike exhibited a westward motion until it reached the periphery of the ridge. It then moved northwestwards and made landfall as category 2 hurricane near Galveston, Texas around 07 UTC 13 September 2008. The system caused about 100 fatalities and damages in several states of the US and ranges among the 10 costliest hurricanes (Blake and Gibney, 2011). After landfall the system started to recurve and underwent ET, which was completed around 12 UTC 14 September. The extratropical system moved northeastwards and caused hurricane-force wind gusts in northeastern parts of the US. The remnants of Ex-Ike interacted with a mid-latitude ridge, which amplified strongly thereafter. Ahead of this mid-latitude ridge, Ex-Ike moved towards the Denmark Strait where it finally decayed. At the front flank of this ridge, a short-wave trough formed west of Portugal, that became a cut-off low accompanied by a surface cyclone during the next days. In the end, this system moved towards the Iberian Peninsula. Detailed information about Hurricane Ike is provided in the Tropical Cyclone Report by citetBerg2009. Three ensemble forecasts for the ET of Hurricane Ike are included in the TIGGE study (Chapter 4).

2.4.3 Typhoon Sinlaku

A tropical wave in the western North Pacific formed the precursor of Typhoon Sinlaku. After several days, around 00 UTC 8 September 2008, the wave intensified into a tropical depression and reached tropical storm intensity around 18 UTC on that day (Figure 2.7). Typhoon Sinlaku moved north-northwestwards and intensified rapidly between 12 UTC 8 September 2008 and 12 UTC 10 September 2008 in a region of high ocean heat content and low vertical wind shear. After 11 September 2008 the system moved towards the west and weakened slightly before it made landfall in northern Taiwan on 06 UTC 14 September 2008. After crossing Taiwan the weak remnants recurved into the East China Sea, where the cyclone reinten-

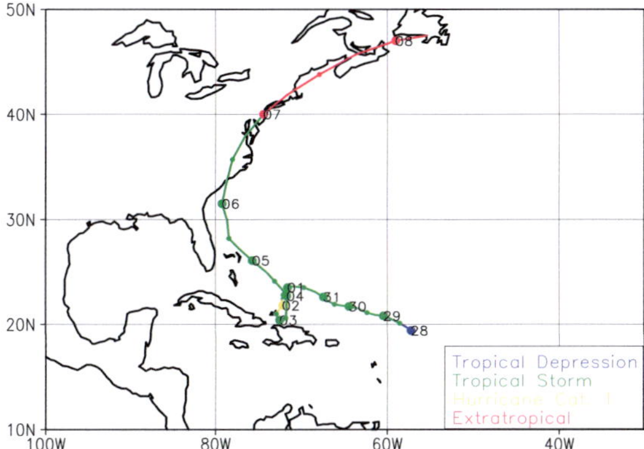

Figure 2.5: Track for Hurricane Hanna in August-September 2008. Dots mark the position of Hurricane Hanna every 6 hours. Large dots describe position at 00 UTC of the indicated day (number). Intensity in colours, as given in legend. Based on best track data from the HURDAT-archive from the National Hurricane Centre (www.nhc.noaa.gov/pastall.shtml#hurdat). ET positions from "Historical Hurricane Tracks" from the National Oceanic Atmospheric Administration (NOAA, http://www.csc.noaa.gov/hurricanes/#).

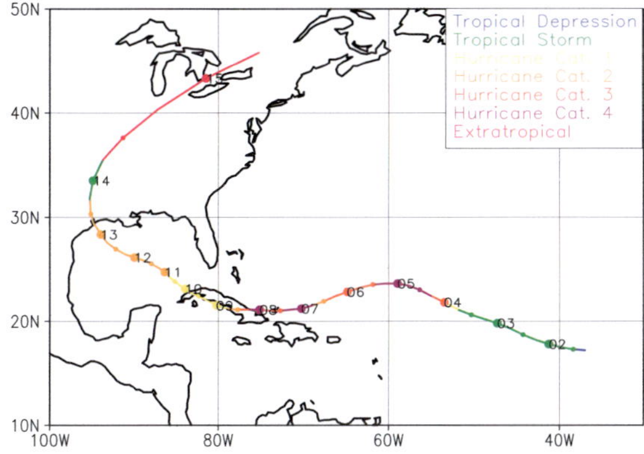

Figure 2.6: As Figure 2.5 but for Hurricane Ike in September 2008.

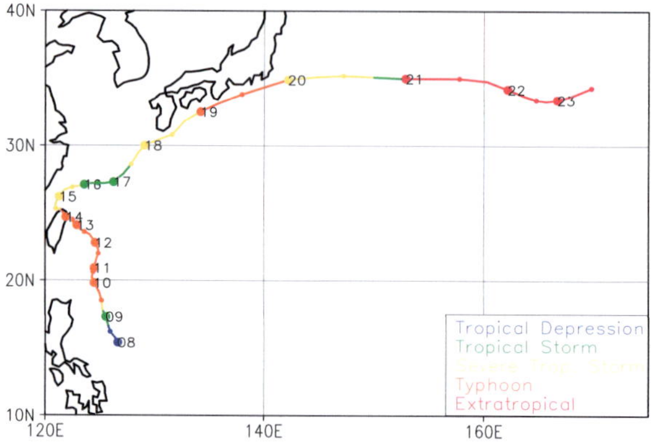

Figure 2.7: Track for Typhoon Sinlaku in September 2008. Dots mark the position of Typhoon Sinlaku every 6 hours. Large dots describe position at 00 UTC of the indicated day (number). Intensity in colours, as given in legend. Based on best track data from the Regional Specialised Meteorological Center (RSMC), Typhoon Center-Tokyo, Japan Meteorological Agency (JMA, http://www.jma.go.jp/jma/jma-eng/jma-center/rsmc-hp-pub-eg/besttrack.html).

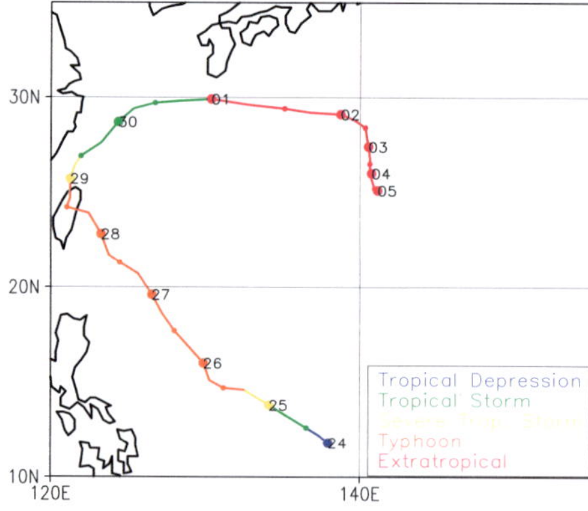

Figure 2.8: As Figure 2.7 but for Typhoon Jangmi in September-October 2008.

sified strongly as a tropical system. The reintensified tropical cyclone then moved along the east coast of Japan. It weakened and developed extratropical characteristics until 21 September 2008. Typhoon Sinlaku occurred during the THORPEX-Pacific Asian Regional Campaign (T-PARC) in the Typhoon Season 2008. During a series of eleven aircraft mission throughout the whole life cycle of Typhoon Sinlaku a large set of aircraft reconnaissance data sets was collected, which enables detailed studies of the different stages in the development of Typhoon Sinlaku to be conducted. This will be of special interest for the period of the reintensification as a tropical cyclone after landfall in Taiwan and the subsequent transformation into an extratropical system near Japan. Some information about Typhoon Sinlaku is provided by Cooper and Falvey (2008) and the in the Annual Report of the Regional Specialised Meteorological Centre, Tokyo JMA (2008). More details about Sinlaku can be found in Sanabia (2010), Förster (2011) and Quinting (2011), who used T-PARC data for a detailed examination of the structure of TY Sinlaku before and during its ET. Three ensemble forecasts for the ET of Typhoon Sinlaku are considered in the TIGGE study.

2.4.4 Typhoon Jangmi

Typhoon Jangmi (Figure 2.8) formed from a tropical depression around 23 September 2008 in the Philippine Sea. Under strong intensification it headed north-northwestwards towards Taiwan. Jangmi became a typhoon at around 06 UTC 25 September 2008. Shortly before landfall in Taiwan at around 09 UTC 8 September 2008, Jangmi was classified as super-typhoon. Strong precipitation and wind gusts caused flooding and power outages in Taiwan. During landfall Jangmi weakened, underwent an abrupt northward deflection and headed towards the East China Sea north of Taiwan. It moved off the coast of Taiwan as a tropical storm. The storm recurved and approached an upper-level jet-streak over Japan. The interaction between the upper level outflow of Jangmi and the jet-streak at around 30 September 2008 accelerated the upper-level winds and thus enhanced the strength of the jet-streak and led to the development of a weak ridge (Grams, 2011). Finally, Jangmi transitioned into an extratropical cyclone, passed the southern tip of Japan and decayed near 140°E around 5 October 2008. T-PARC reconnaissance flights also collected a large amount of data for different stages in the life cycle of Jangmi. A brief overview about Jangmi is provided in the Annual Report of the Regional Specialised Meteorological Centre, Tokyo JMA (2008). More details on the interaction between Jangmi and the mid-latitude flow can be found by Grams (2011). One ensemble forecast for the ET of Typhoon Jangmi is considered in the TIGGE study (Chapter 4).

2.4.5 Tropical Storm Bavi

Tropical Storm Bavi (Figure 2.9) developed northeast of Guam as a region of deep convection at around 12 UTC 18 October 2008 and was declared as tropical storm by JMA at 06 UTC 19 October 2008. Shortly after this time, Bavi underwent recurvature. Bavi accelerated ahead of a weak upper-level trough and transitioned subsequently into an extratropical system. In the following hours, the mid-latitude wave pattern amplified strongly and favoured the further intensification of the transitioning storm. By 06 UTC 22 October 2008 the central pressure of the extratropical system had decreased to 948 hPa and hurricane-force winds were recorded. Bavi intensified further until 00 UTC 23 October and moved towards the northeast. By 12 UTC 24 October 2008 Ex-Bavi reached the coast of Alaska as a strong extratropical system ahead of a tilted upper-level trough. A brief overview of Bavi is given by the Annual Report of the Regional Specialised Meteorological Centre, Tokyo JMA (2008). Two ensemble forecasts for the ET of Tropical Storm Bavi are incorporated in the TIGGE study.

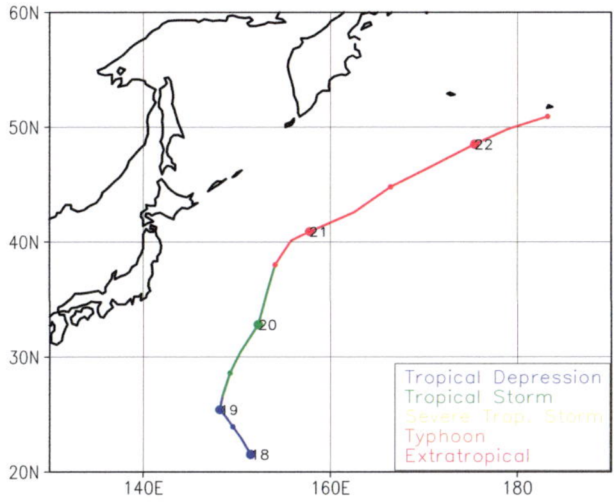

Figure 2.9: As Figure 2.7 but for Tropical Storm Bavi in October 2008.

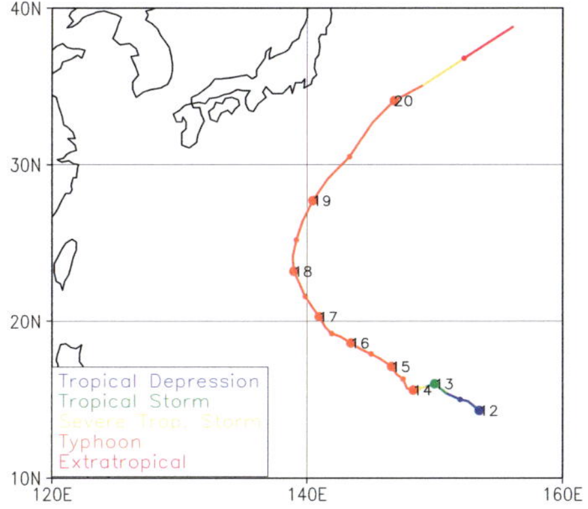

Figure 2.10: As Figure 2.7 but for Typhoon Choi-Wan in September 2009.

2.4.6 Typhoon Choi-Wan

Typhoon Choi-Wan emanated from a tropical depression which formed around 00 UTC 12 September 2009 (Figure 2.10). During its west-north-westward movement on the first day, the system intensified and was classified as tropical storm at around 18 UTC. Under further intensification it moved predominantly towards the west, became a typhoon at around 00 UTC 14 September 2009. During the subsequent days the system intensified further and reached its maximum intensity on 16 September 2009 while following a northwestward track. On 18 September, Choi-Wan started to recurve and accelerated towards the mid-latitudes. It moved ahead of a preexisting mid-latitude trough and an associated extratropical cyclone. Around 21 September the remnants of Choi-Wan merged with the mid-latitude system and the merger reintensified strongly as an extratropical cyclone. As Choi-Wan remained over the ocean during its whole life-cycle, it had an overall minor direct impact. During ET Choi-Wan approached a preexisting moderate Rossby wave pattern in the mid-latitudes. The interaction between the transitioning storm and the mid-latitude flow led to an amplification of the mid-latitude wave pattern over the entire North Pacific. Wave breaking over North America formed a cut-off cyclone east of the Rocky Mountains. The circulation in the narrowed trough and the cut-off was associated with a temperature drop of about 15-20 K in the western parts of the Great Plains between 21 and 25 September 2009. More general information about Choi-Wan is provided in the Annual Report of the Regional Specialised Meteorological Centre, Tokyo JMA (2009). Grams (2011) examined the ET of Choi-Wan and its impact on the downstream flow structure in detail. In the present study several ECMWF ensemble forecast scenarios for the ET of Choi-Wan are investigated using an analysis of the eddy kinetic energy budget.

3 Methodologies

The core of this work is the investigation of ensemble forecasts for ET events under the application of different analysis approaches. The techniques which are employed in this thesis are presented in this section. The first method to describe is the analysis of the Empirical Orthogonal Functions (EOF) and the subsequent fuzzy clustering. This method enables ensemble members with related synoptic developments to be grouped together in order to achieve an overview of the main possible development scenarios contained in the ensemble forecast in question. The extracted development scenarios then facilitate the investigation of the physical processes causing these different scenarios and the identification of the contribution of the storm or other synoptic features to the observed scenarios. A means for this closer examination is provided by analysing the eddy kinetic energy budget of the individual scenarios. This budget furnishes insight into the energetics of the different development scenarios against the background of a mean synoptic situation and is introduced in the second part of this section,

3.1 Investigation and Simplification of Ensemble Forecasts

The TIGGE ensemble forecasts investigated here contain 231 ensemble members and thus 231 different possible development scenarios. These are too many to be investigated individually. This necessitates a way to condense the bulk of information furnished by TIGGE without losing important information of the variability expressed by the original data set. In terms of ensemble forecasts such a way should enable to summarise the information provided by the 231 individual members without losing the variability they express. In particular this variability, i.e. the dominant differences in their development scenarios, is the desired information.

3.1.1 Empirical Orthogonal Function Analysis

One possibility for such an analysis in atmospheric sciences was pioneered and introduced as Empirical Orthogonal Function (EOF-) analysis by Lorenz (1956), although others had already applied related methods some years before (cf. Hannachi et al., 2007). The basic idea behind this method is that the atmospheric fields under investigation often contain redundant information (Wilks, 1995). In the context of the 231 individual ensemble members, there exist regions where all members show nearly identical fields, i.e. low geopotential heights towards the north or a nearly identical wave pattern in the regions of low forecast uncertainty. Under this assumption of redundant information, the EOF-analysis, also often referred to as Principal Component analysis (PCA, Björnsson and Venegas, 1997) then allows to define a new set of variables that permits the expression of large parts of the variance stated by the original data set at one specific time of investigation, but has a lower dimensionality. This new set of variables is achieved by searching for a new set of vectors $\mathbf{e_i}$ which best describes all synoptic patterns in the ensemble members simultaneously (Kutzbach, 1967). In other words, one has to identify those vectors $\mathbf{e_i}$ onto which the projection of the original data matrix $\mathbf{X}$ explains the maximum variability (Hannachi et al., 2007). Thereby, the $M \times P$ data matrix $\mathbf{X}$ contains the synoptic field under investigation in each of the M ensemble members, discretised at P grid points at one specific time. As we will see later, it makes sense to first compute the anomaly matrix, which contains the deviations of the M ensemble members from the ensemble mean at the P grid points, i.e.

$$\mathbf{X}' = \mathbf{X} - \frac{1}{P}\mathbf{I}_p\bar{\mathbf{x}} \tag{3.1}$$

with $\mathbf{I}_p$ being the identity matrix, which transfers the mean state vector $\bar{\mathbf{x}}$ into matrix dimensions. This leads to the anomaly matrix $\mathbf{X}'$, serving as starting point for the remaining analysis

$$\mathbf{X}' = \begin{pmatrix} x'_{11} & x'_{21} & \cdot & \cdot & \cdot & x'_{P1} \\ x'_{12} & & & & & \cdot \\ \cdot & & & & & \cdot \\ \cdot & & & & & \cdot \\ x'_{1M} & & \cdot & \cdot & \cdot & x'_{PM} \end{pmatrix} \tag{3.2}$$

$$\overset{\text{number of grid points}}{\longrightarrow} \qquad \Big\downarrow \text{\textit{number of members}}$$

Coming back to the original problem, to find the direction of maximum variability, one has to maximise the expression $VAR(\mathbf{X}'\mathbf{e})$, which can be written as (Hannachi, 2004)

$$VAR(\mathbf{X}'\mathbf{e}) = \frac{1}{n-1}\|\mathbf{X}'\mathbf{e}\|^2 = \frac{1}{n-1}(\mathbf{X}'\mathbf{e})^T(\mathbf{X}'\mathbf{e}) = \mathbf{e}^T\mathbf{S}\mathbf{e}$$

with $\mathbf{S}$ being the covariance matrix of the anomaly matrix $\mathbf{X}'$. Under the constraint that $\mathbf{e}$ has unit-length, the maximisation problem becomes bounded (Hannachi, 2004)

$$\max_a(\mathbf{e}^T\mathbf{S}\mathbf{e}), \quad \text{s.t. } \mathbf{e}^T\mathbf{e} = 1 \tag{3.3}$$

whereas the transformation (due to the projection) remains orthogonal and the scales are conserved in both systems (Böker, 2005). Finally, the maximisation of Equation 3.3, using the Lagrange multiplier λ requires the solution of the eigenvalue problem for the covariance matrix $\mathbf{S}$ of the data set

$$\mathbf{S}\mathbf{e} = \lambda\mathbf{e} \tag{3.4}$$

wherein $\mathbf{e}$ and λ are now referred to as the eigenvector and its corresponding eigenvalue of the covariance matrix $\mathbf{S}$ (Kutzbach, 1967). These eigenvectors are the desired Empirical Orthogonal Functions[1] we are looking for. There exist as many eigenvectors $\mathbf{e}_i$ with $i = 1,...P$ as there are grid points in the original data (von Storch and Navarra, 1995), indicating the the direction of the maximum joint variability in the data set under the constraint that they are orthogonal to each other. Thus, for all eigenvectors, Equation 3.4 yields

$$\mathbf{S}\mathbf{E} = \mathbf{E}\mathbf{L} \tag{3.5}$$

with $\mathbf{E}$ being a PxP matrix that contains all existing eigenvectors as columns and the PxP diagonal matrix $\mathbf{L}$ whose i-th diagonal element λ_i is the corresponding eigenvalue. According to custom, the eigenvectors and eigenvalues are sorted in decreasing order (e.g. von Storch and Zwiers, 1999, Hannachi et al., 2007). Thus, the first eigenvector is associated with the strongest variability in the data set and points in that direction, where the ensemble members show the strongest variance. The second eigenvector points in the direction where the ensemble members have the second strongest variability, under the constraint that it has to be orthogonal to the first eigenvector, and so on. In other words, the eigenvectors span a new orthogonal coordinate system to view the data, whose axes are aligned parallel to the directions where the data exhibit maximum variability (Wilks,

---

[1] EOFs are defined empirically for the data set of interest and are different for varying data sets, what distinguishes the EOFs from other theoretical orthogonal functions, which fit, once determined, for all data sets (Wilks, 1995)

1995). This requirement of orthogonality causes the EOFs to be uncorrelated in space (Björnsson and Venegas, 1997).

As the anomaly matrix acted as starting point for the analysis, the magnitude of the eigenvalues λ_i now directly indicates the amount of variance explained by the i-th eigenvector or EOF $\mathbf{e}_i$, with $i = 1,...,P$ (Kutzbach, 1967). The fraction of the total variance in the anomaly field, which is captured by the i-th eigenvalue can be expressed as percentage value (Hannachi, 2004)

$$\frac{100\lambda_i}{\sum_{i=1}^{P}\lambda_i}\% \tag{3.6}$$

Often large parts of the joint variability are explained by the first few eigenvectors, which then clearly dominate the remaining ones (von Storch and Zwiers, 1999). Furthermore, according to Lorenz (1956), eigenvectors that describe large parts of the variability predominantly capture real large scale physical features, i.e. distinct variabilities of the large scale flow. Such features are intended to be extracted with this method. On the other hand, the variability captured by the eigenvectors with small eigenvalues is mainly due to small scale disturbances in the synoptic pattern, or is caused by measurement and model uncertainties. By retaining only the first few dominant eigenvectors, a large amount of the variance in the original data set can now be expressed by just a few new base vectors, and the goal of decreasing the dimensionality of the data set while retaining its dominant variance is achieved. As the EOFs describe where the atmospheric fields under investigation show strongest differences, they generally allow to identify typical modes of variability in the atmosphere, such as large scale circulation patterns (Harr and Elsberry, 1995, Cheng and Wallace, 1991), variation patterns of climate variables (Kutzbach, 1967, Kalkstein et al., 1987) or more specific features like the Arctic Oscillation (Hannachi et al., 2007). In the case of ensemble forecasts the EOFs capture those synoptic features of the flow fields that are represented differently in the individual members.

The contribution of each ensemble member to the extracted patterns of variability is stated by its principal components or PCs. The PCs $\mathbf{c}$ of the M ensemble members are obtained by mapping the original data $\mathbf{X}$ onto the new coordinate system $\mathbf{e}$

$$\mathbf{c} = [\mathbf{E}]^T \mathbf{X} \tag{3.7}$$

or for the projection of the m-th member on the i-th EOF, with p running over all grid points

$$c_i(m) = \sum_{p=1}^{P} x(m,p)e_i(p) \tag{3.8}$$

For this reason the PCs are nothing other than the coordinates of the M ensemble members in the new P-dimensional coordinate system. In their investigation of ECMWF ensemble forecasts during ET events, Harr et al. (2008) and Anwender et al. (2008) found large parts of the variability (up to 40%) and clear variation patterns of the atmospheric flow fields to be described by the two dominant EOFs (EOF1 and EOF2). Hence, the focus in this study is put on those two dominant EOFs and the corresponding PCs as well. In this way, all ensemble members can be expressed by their PC1-PC2 pair in the 2-dimensional EOF phase space. Thus, instead of having M ensemble members with P grid points, the problem is now condensed to one point for each of the M members in the 2-dimensional PC phase space. However, if this limitation to only the first two EOFs becomes equivocal during the analysis (i.e. due to large variability or significant variation patterns expressed by EOF3) the method can be extended to account for more EOFs as well.

The principal components express the contribution of the individual members to the new coordinate system. Thus, members with related PCs contribute similarly to the pattern of variability and show analogue development scenarios. This fact is utilised in the second part of the analysis technique.

3.1.2 Fuzzy Clustering

Once the PCs for the individual ensemble members are defined, they can be used as representatives for the members in the phase space spanned by the first two EOFs. As members with related PCs show related development scenarios, the PCs are now used as input variables for a clustering procedure to separate the ensemble members into different groups.

Clustering procedures are a well known method to separate large data sets into several groups and are used for research purposes in natural- as well as social- and economic sciences (Pang-Ning Tan, 2005, Anderberg, 1973). Due to the variety of problems that can be addressed with a cluster analysis, there exist several approaches which show distinct performance on different data sets. Overall, the methods can be separated in two groups: hierarchical- and non-hierarchical or partitional clustering. The hierarchical methods are also referred to as nested methods (Anderberg, 1973, Pang-Ning Tan, 2005), as they allow the existence of sub-clusters. A hierarchical method leads to the development of a tree-like structure or dentagram (Wilks, 1995), which describes the relation between the individual members. The most common way of hierarchical clustering, the agglomerative way, starts with as many clusters as there are data points, i.e. members. During the clustering procedure these clusters are merged step by step, based on their similarities, expressed e.g. by their Euclidean distances. Finally, all members are assigned to one cluster (the root of the tree Anderberg, 1973). This results in as many cluster solutions as there are members or data points. Out of these M possible clusters, where M is the number of members, one has to chose the optimum number of clusters (K) for the data set. One possibility is to use a certain threshold for the distance between two clusters, which they have not to exceed if they should be merged or divided in the next step. The tree-like structure also implies that the data points, once they are separated or merged, stay in this relation (Anderberg, 1973, Ketchen and Shook, 1996). Thus, the individual members cannot be re-assigned to another cluster which might evolve during the hierarchical clustering, although they might be related to another cluster as well. This strict assignment to one cluster, referred to as hard clustering, also reveals that the members within one cluster have nothing in common with members from other clusters (Bezdek et al., 1984).

Non-hierarchical or partitioning clustering algorithms separate the members directly into a predefined number of K clusters, based on their distances to the cluster centres. The method starts with initially chosen seed points and assigns the data points to these cluster centres, based on their Euclidean distance. Then, in an iterative procedure, the cluster centres are re-computed, followed by a re-assignment of the data points, which means that the data points may change their cluster assignments. This iterative procedure stops if no members change clusters anymore, and thus the cluster centres remain the same from one iteration to the next (Anderberg, 1973). Basically, a non-hierarchical clustering also leads to a clear- or hard assignment of the individual data points to only one of these clusters. However, this method can be adapted to overcome hard assignment (Bezdek et al., 1984). Against the background that physical processes might not have clear assignments to unique clusters[2], Zadeh (1965) introduced the idea of fuzzy sets, which basically allows the members to contribute, i.e. have a membership to more than one cluster. Thereby, this membership is proportional to the distance between member and cluster centre. Based

---

[2]like e.g. fruits, which can be grouped in apples and cherries

thereon, Bezdek (1981) developed a non-hierarchical fuzzy c-means clustering algorithm. Mo and Ghil (1988) adapted the fuzzy clustering in such a manner that they also allow members to be unassigned to clusters if they are located in border regions and indicate related memberships to two or more clusters. Thus, members whose memberships are below a distinct threshold are not assigned to the cluster. For the PCs or development scenarios this means that these members show a synoptic pattern which lies in between the main pattern identified by the clusters. This fuzzy-clustering approach was also proposed by Cheng and Wallace (1991) to overcome unstable cluster assignments in several realisations of their hierarchical clustering for the 500 hPa geopotential height field.

The main advantage of such a non-hierarchical clustering procedure is the possibility for members to change clusters. This possibility, together with the iterative procedure of re-computing the cluster centres, allows to maximise the in-cluster similarity during the cluster procedure, while the in-between cluster similarity is minimised (Bezdek et al., 1984). The disadvantage of the non-hierarchical clustering is that the number of clusters must be defined in advance. One possibility to resolve this issue, which is also applied in this study, is to repeat the analysis with different numbers of clusters and then identify the optimum number of clusters. How the number of clusters is defined is described in Section 3.1.3. In this study, a fuzzy c-means clustering is applied to the PCs of the ensemble members. This method, originally developed by Harr and Elsberry (1995), was already applied successfully to extract the main development scenarios in forecasts of ET events by Harr et al. (2008) and Anwender et al. (2008).

Mathematically, the method can be described according to Bezdek (1981). In our specific case, we have a set of M=231 members or data points, given in the 2-dimensional PC phase space $\mathbf{y}=(PC1_1, PC2_1, ..., PC1_M, PC2_M)$. With the fuzzy clustering algorithm, we want to divide this data set into K subsets $\mathbf{y}_i$. This results in a $K \times M$ matrix $\mathbf{U}$ that contains the membership

$u_{mk} \in [0,1]$ of each of the M members to every of the K clusters, based on two assumptions:

$$\sum_{m=1}^{M} u_{mk} > 0 \quad \text{for all } m \rightarrow \text{(each member has a membership to every cluster);}$$

$$\sum_{m=1}^{M} u_{mk} = 1 \quad \text{for all } k \rightarrow \text{(memberships for one member sum up to 1);}$$

The membership matrix $\mathbf{U}$ is gained by minimising a generalised least-squared errors functional $J(U,v)$ (Bezdek, 1981) in an iterative process

$$J(U,v) = \sum_{m=1}^{M} \sum_{k=1}^{K} u_{km}^{f} d_{km}^{2} \tag{3.9}$$

with $d_{km}^{2} = (y_m - v_k)^{T}(y_m - v_k)$ being the euclidean distances between the member y_m and the cluster centre v_k. The fuzzifier exponent f influences the strength of assignment between the members and the clusters, whereas $f{=}1$ stands for the strongest assignment and typically values of f between 1 and 3 are suitable (Bezdek et al., 1984). In our case, $f{=}1.5$ is used according to Harr et al. (2008). By minimising Equation 3.9 the new cluster centres v_k and the elements of the new membership matrix $\mathbf{U}_{1+i}$ can then be computed as (Bezdek, 1981)

$$v_k = \frac{\sum\limits_{m=1}^{M} u_{km}^{f} y_m}{\sum\limits_{m=1}^{M} u_{km}^{f}} \tag{3.10a}$$

$$u_{km} = \frac{1}{\sum\limits_{k=1}^{K} \left(\dfrac{d_{km}}{d_{jm}}\right)^{2/(f-1)}} \tag{3.10b}$$

where d_{km} stands for the euclidean distance between member m and the cluster centre in question (cluster k) and d_{jm} represents the euclidean distance between member m and each other cluster centre j ($j \neq k$).

At the beginning of this iterative procedure (iteration step $i = 1$), K (as many clusters as necessary) pairs of PCs are randomly selected to form initial cluster centres v_1 for the clustering algorithm. The initial membership matrix U_1 is then defined by Equation 3.10b, depending on the euclidean distance between the M members and the K cluster centres. Then, Equation 3.10a furnishes the new cluster centres v_2. Re-assignment of the members to these new centres then leads to the new membership matrix U_2. This iterative procedure of re-computing the cluster centroids and the membership matrix is repeated, until the differences between the old and new membership matrix $U_{i+1} - U_i$ is less than $10e^{-5}$. In this way, the centroids are placed in the optimum position. At the end of the clustering procedure, the membership-matrix U finally contains the weight w, with which each member is assigned to each cluster. The difference between the mean of the strongest weight $\overline{w_{max}}$ for each member and the standard deviation of these dominant weights $\sigma_{w_{max}}$, $thres = \overline{w_{max}} - \sigma_w$ is used as a threshold to define whether a member is assigned to a cluster or not. Members, whose weights are below the threshold ($w < thres$) are not assigned to any cluster, as they lay in between two clusters and have related assignments to both of them.

The result of the fuzzy clustering depends on the initially chosen seed points, which means that the outcome of the clustering may vary for a distinct set of initial seed points. Therefore, the method of Harr et al. (2008) was adapted to meet this concern. For each pre-defined number of clusters, the clustering algorithm is now applied 100 times to the data set which leads to 100 realisations of the clustering. These realisations are then displayed in terms of their centroids in the PC1/PC2 phase space. If the clustering of the members is independent of the chosen seed points the 100 realisations show the same cluster solution and the cluster centroids are located at the same place in all 100 solutions. This can also be seen as a reproducibility of the cluster solution (Christiansen, 2007), and is referred to as a stable clustering solution in the remainder. For our purposes, only the stable solutions are then taken into account to define the optimum number of clusters.

3.1.3 How Many Clusters Should be Retained?

The question of how many clusters should be allowed in the clustering procedure is known to be difficult to answer and is discussed extensively in the literature (e.g. Milligan and Cooper, 1985, Christiansen, 2007). For this study, an adequate and rudimentary objective method for defining the optimum number of clusters is employed to make the outcome of the different cases under investigation as comparable as possible. The clustering procedure is applied to the data set multiple times with different numbers of clusters, typically 2 to 8. Only those numbers of clusters that produce a stable solution are then taken into account for selecting the optimum number (Arnott et al., 2004). For those stable solutions, the scatter plots of the clusters are investigated. They exhibit the normalised distribution of the members in the PC1/PC2 phase space and show the separation of the clusters. If, for example, the clusters are well separated in the solution with 5 clusters, but in the solution with 6 clusters a small amount of outliers start to separate from the main solutions (Figure 3.1, top), then 5 clusters are chosen as the optimum number. This addresses the concern of extracting the main possible development scenarios instead of some outliers. As the outliers are contained in the data set anyway, they could be investigated manually. Another criteria to stop the clustering procedure is reached if the new cluster occurs to be "squeezed-in" between preexisting clusters (Figure 3.1, bottom). In this case, the basic assumption of the clustering to maintain largest possible differences between the individual clusters is violated. In terms of the development scenarios, this squeezed-in cluster is often resembled by the preexisting clusters. To identify such relations, the mean development scenarios stated by the individual members are taken into account as well. This solution is chosen only if the new cluster solution leads to a clearly distinct development scenario compared to the former ones. Otherwise, the former number is used for the remainder of the analysis.

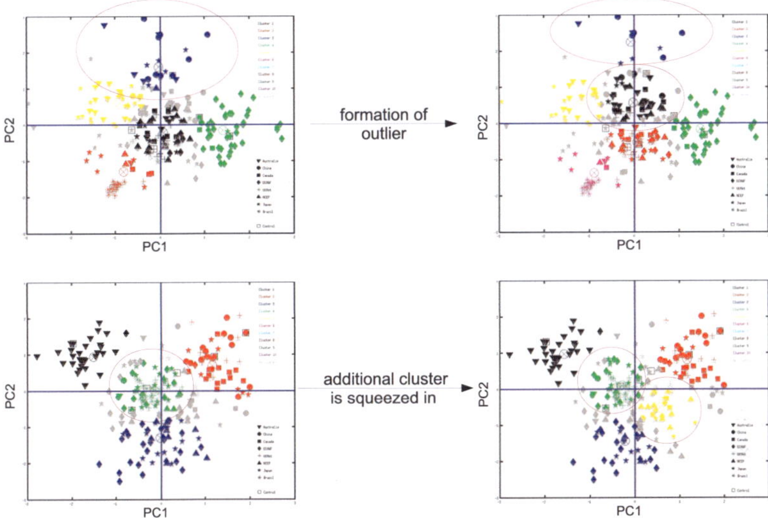

Figure 3.1: Example for undesirable cluster solutions. Top: increase of cluster number from 5 to 6 causes the additional cluster to form an outlier. Bottom: increase of cluster number from 4 to 5 leads the additional cluster to be squeezed-in between the preexisting clusters. If such a partitioning occurred, the former number of clusters (top: 4, bottom: 5) is retained.

3.1.4 Application of the EOF- and Fuzzy Clustering Analysis

With the analysis technique specified above, the bulk of information which is furnished by TIGGE or another EPS can be condensed to a few case-dependent number of clusters. These clusters or subsets of the ensemble system then capture the main possible development scenarios contained in the forecast under investigation. Hence, they provide the background for further investigations of the physical processes that cause the differences among the observed scenarios using other analysis techniques. Therefore, the development scenarios can mainly be expressed in two ways. On the one hand, one can build the mean from all members in the cluster and examine this cluster mean state in further detail. This might take into account some extreme cases (deep trough or strongly amplified ridge) as well, as they are considered and influence the mean, but it will also smooth out some details (short wave trough with slightly different positions in cluster members) in the development. On the other hand, one might select a representative member from the cluster, which best resembles the cluster mean. Thereby, no details would be smoothed out due to averaging, but no extreme developments would be considered. This representative member is then chosen to be the member with closest Euclidean distance to the cluster centre in the two-dimensional PC1-PC2 phase space. In the study at hand, EOF- and fuzzy clustering analysis is mainly applied to the geopotential height field at 500 hPa, as this variable nicely captures characteristics of the mid-latitude flow as well as the remnants of the transitioning tropical cyclone. Another approach in using this analysis technique is undertaken in Chapter 5. The specific forecast time at which the analysis procedure is applied to the data, is defined in dependence of an increase in forecast uncertainty (see Chapter 4 for more details).

The description of the analysis method discussed above was tailored to the use of the TIGGE multi-model EPS with its 231 ensemble members. The method can also be applied to subsets of the TIGGE EPS or individual EPS, like the ECMWF or NCEP EPS, as it was done by Anwender et al. (2008) and Harr et al. (2008).

3.1.5 Tools to Examine Cluster Results

During this work we will encounter some questions about the differences or similarities among distinct cluster solutions or development scenarios. Therefore, two convenient tools to examine cluster results are introduced in the following.

a) Rand Index

A measure for the similarity between two cluster solutions is provided by the Rand Index *RI* (Rand, 1971). This index is very helpful to determine whether the cluster solution for one forecast variable separates the members in the same way as the cluster solution for another forecast variable does. If we apply two distinct clustering approaches (e.g. using 500 hPa geopotential height and 850 hPa temperature as variables for EOF- and cluster-analysis, respectively) to an ensemble with, for example, $N = 6$ members, we will get two distinct solutions Y and Y' on how to group the ensemble members. In this example, both clustering approaches will result in two clusters, $Y_{1,2}$ for using the one and $Y'_{1,2}$ for using the other forecast variable. Thereby, the $N = 6$ ensemble members are assigned to the clusters as follows.

$$Y_1 = 1,2,5 \quad Y_2 = 3,4,6 \quad \text{and} \quad Y'_1 = 4,6 \quad Y'_2 = 1,2,3,5$$

The Rand Index now simply counts the number of members that are assigned together or differently in the two cluster solutions and that are mixed, i.e assigned together in one, but differently in the other solution. For the example stated above (adapted from Rand, 1971), we find the following result

Member pair	Together	Separate	Different
1&2	*		
1&3			*
1&4		*	
1&5	*		
1&6		*	
2&3			*
2&4		*	
2&5	*		
2&6		*	
3&4			*
3&5			*
3&6			*
4&5		*	
4&6	*		
5&6		*	

Thus, ten member-pairs are assigned in the same way in the two solutions (together in both or separate in both), while five member-pairs are assigned differently. Dividing the amount of members assigned to the same solutions by the number of possible solutions for assignment in the two approaches then leads to the Rand Index. In summary, this can be written as (Rand, 1971)

$$RI(Y,Y') = \frac{\sum_{i<j}^{N} \gamma_{ij}}{\binom{N}{2}} \tag{3.11}$$

where N stands for the number of members, and X_n and γ_{ij} are defined to be

$$\gamma_{ij} = \begin{cases} 1 \text{ if } X_i \text{ and } X_j \text{ are assigned together in both} \\ 1 \text{ if } X_i \text{ and } X_j \text{ are assigned separated in both} \\ 0 \text{ if } X_i \text{ and } X_j \text{ are are together in one but separate in the other} \end{cases}$$

Thus, RI varies between 1, if the two cluster solutions are identical, and 0, if none of the members are assigned in the same manner in the two solutions. With $RI = 0.67$, the example above would be referred to as a rather moderate agreement.

b) Similarity Index

For comparing different development scenarios or atmospheric fields, an objective measure for similarity is desirable. Based on an idea of Buizza (1994), a similarity index SI is defined, which provides information about the similarity of two fields under investigation. Therefore, the atmospheric developments are represented as state vectors in the phase space of possible synoptic developments. Then, their similarity can be expressed as the mean absolute value of the dot product relationship between these two state vectors, normalised by the product of the mean absolute values of the two fields under investigation according to

$$SI = \frac{|\text{field}_1 \cdot \text{field}_2|}{|\text{field}_1| \, |\text{field}_2|} \tag{3.12}$$

SI varies between 1 for equal fields and 0 for fields that have nothing in common, or in other words are orthogonal in phase space of possible synoptic developments. On the one hand, SI is used to explore similarities among EOF distributions found in distinct data sets (Chapter 4). On the other hand, it will be used as a measure for analogy among development scenarios. For the latter application it makes sense to first compute a mean development of both fields under investigation. Then the perturbations of the data fields from this mean state are computed before SI is assigned. Otherwise, as large parts of the fields are often similar[3], SI will be close to 1, although the representation in distinct regions is different. Furthermore it can be helpful to restrict the computation of SI only on that regions of the fields, which show strong differences. However, as the SI is also sensible to differences in magnitude of the perturbations, there might exist potential for improvement in future applications.

---

[3]two geopotential height fields for the same northern hemispheric domain will always coincide in having low heights to the north and increased heights in the south

3.2 An Eddy Kinetic Energy Approach to Examine Different Development Scenarios

The individual development scenarios resulting from the fuzzy clustering of the ensemble forecast now provide the means for detailed studies of the underlying physical processes. To increase the understanding about how the ET process may perturb the mid-latitude circulation and thus the synoptic development in downstream regions, Harr and Dea (2009) investigated the ET of four tropical cyclones, using a downstream baroclinic development paradigm, which was developed by Orlanski and Sheldon (1995). In this paradigm, which will be discussed in detail below, the cyclones and wave patterns in the mid-latitude flow are highlighted as "eddies", i.e. deviations from the mean circulation pattern in the mid-latitudes which show enhanced kinetic energy, compared to a mean kinetic energy distribution. By analysing the kinetic energy budget, the contribution of the distinct flow features to the development in downstream regions can be investigated. Motivated by the findings of Harr and Dea (2009), eddy kinetic energy analysis is now employed to study physical processes in the distinct development scenarios found in ensemble forecasts. After a brief introduction into the energetics of the atmosphere and the background information for the eddy kinetic energy budget analysis will now be introduced and explained in detail.

3.2.1 Energetics of Atmospheric Circulations

Incoming short-wave solar radiation and outgoing long-wave radiation balance the energy budget of the earth-atmosphere system. Differential heating in tropical and polar regions, together with the rotation of the earth establish a complex circulation pattern in the global atmosphere. Thereby, the Coriolis force inhibits a direct thermal exchange between polar and equatorial regions. The mid-latitude meridional temperature gradient increases, which in turn enhances baroclinicity in the extratropical flow. Instabilities in the extratropical flow then trigger the formation of a mid-latitude wave pattern and associated cyclonic and anticyclonic systems which are responsible for the temperature and energy exchange in the mid-latitudes. Barotropic instability thereby supports the development of upper-level wavelike disturbances on a sloping tropopause (Rossby, 1940) and explains the formation of mid-latitude wave patterns. During their intensification, barotropic instabilities extract kinetic energy from the mean flow. Distinct development theories exist about the formation and amplification of baroclinic extratropical cyclones. In general, these cyclonic systems are initialised by small disturbances in a baroclinic environment. Such small scale disturbances are for example single normal mode disturbances which grow under baroclinic instability (normal modes approach Charney, 1947, Eady, 1949), or more complex small scale initial perturbations, which may also include non-normal modes (initial value approach Farrell, 1982, 1984). Such complex initial value perturbations and can grow even in regions with weak baroclinicity, where baroclinic instability hardly may trigger the amplification of disturbances. This initial value approach better resembles cyclogenesis under real atmospheric conditions (Farrell, 1985, 1989, Reed, 1990). Baroclinic instabilities mainly grow by the conversion of available potential (see below) into kinetic energy.

Seen in an energy framework, the differential heating in polar and equatorial regions acts to make parts of the total potential energy of the atmosphere available for conversion into kinetic energy, thus it generates so-called available potential energy (Lorenz, 1955). Total potential energy describes the sum of potential and internal energy of the atmospheric system, as these two quantities are mutually dependent. A minimum of total potential energy is reached, if the whole atmosphere is brought into hydrostatic equilibrium by adiabatic redistribution of mass (cf. Holton, 2004).

Differential heating now causes a distinct increase in total potential energy in equatorial and polar regions, which also disturbs the global hydrostatic equilibrium. The available potential energy is defined as the difference between the actual total potential energy of the atmosphere and the minimum total potential energy that can be reached by an adiabatic redistribution of mass. Only this small fraction ($\sim$0.5%) of the total potential energy is available for conversion into kinetic energy and thus for the generation of atmospheric motions. From this available potential energy just approximately 10% is converted into kinetic energy (cf. Holton, 2004). Parts of the redistributive atmospheric motions are organised as direct thermal circulation cells in the polar and tropical regions, while in the mid-latitudes the cyclonic and anticyclonic systems and their associated wave patterns in the upper troposphere act as thermal redistributors. How these synoptic "disturbances" in the mean zonal flow are established and amplified, i.e. how available potential energy (P) is converted into kinetic energy (K) can now schematically be summarised in terms of the energy cycle of Lorenz (1955) (after Holton, 2004):

1. Differential net heating in polar and equatorial regions creates mean zonal available potential energy

2. Extratropical cyclones and anticyclones cause thermal exchange between warm equatorial and cold polar regions, which transforms mean available potential energy into available potential energy of these cyclonic disturbances

3. Vertical motion in baroclinic disturbances then transforms available potential energy of the disturbances into kinetic energy of the disturbances

4. Cyclolytical processes and poleward transport of angular momentum (cf. Pichler, 1997) transform kinetic energy of the disturbances into kinetic energy of zonal mean and thus help to sustain the mean flow pattern

5. Energy dissipates and enhances the total potential energy due to frictional processes

3.2.2 The Eddy Kinetic Energy Framework

In the early 1990's Isidoro Orlanski and co-authors (Orlanski and Katzfey, 1991, Orlanski and Chang, 1993, Orlanski and Sheldon, 1993, Chang, 1993, Chang and Orlanski, 1993, Orlanski and Sheldon, 1995) could highlight the importance of localised kinetic energy fluxes between adjacent baroclinic systems to cyclone growth and downstream development, i.e. development of baroclinic cyclones downstream of a preexisting mature cyclone. As the basic assumption of these studies is the separation of the atmospheric flow field, and thus its energetics as well, in a time-mean (zonal flow) and a perturbation or "eddy" part, this analysis technique can also be referred to eddy kinetic energy analysis. This coincides with the assumptions made above by discussing the energetics of the atmospheric circulation, but the disturbances are now referred to as eddies.

a) Derivation of Eddy Kinetic Energy Equations

An equation for the budget of changes in kinetic energy can simply be achieved in general, by multiplying the momentum equation per unit mass by the horizontal velocity vector $\mathbf{v} = u\mathbf{i} + v\mathbf{j}$. Therein, the kinetic energy per unit mass is expressed as $K = (u^2 + v^2)/2$ and has units of joules per kilogramme (J/kg). As the model data sets, employed in this study, are interpolated onto pressure levels it is reasonable to use the horizontal, hydrostatic momentum equation per unit mass in pressure coordinates (cf. Haltiner and Williams, 1979)

$$\frac{\partial \mathbf{v}}{\partial t} + \mathbf{v} \cdot \nabla_p \mathbf{v} + \omega \frac{\partial \mathbf{v}}{\partial p} = -\nabla_p \phi - f\mathbf{k} \times \mathbf{v} + \mathbf{F_r} \qquad (3.13)$$

where notation is as common: $\mathbf{v}$ stands for the horizontal and ω for the vertical velocity, ϕ describes the geopotential while $\mathbf{F_r}$ is representing frictional forces. As stated above, we are now interested in the deviation of the kinetic energy from the monthly time-mean, referred to as eddy kinetic energy or K_e. According to the derivation of the set of kinetic energy budget equations by Orlanski and Katzfey (1991) we first have to partition the momentum equation into a time independent mean- and an eddy-part by substituting the horizontal velocity ($\mathbf{v}$) and geopotential (ϕ) in the momentum equation as follows. Thereby, the decomposition of variables and averaging of terms is formally identical to the well-known Reynolds method usually employed in the description of turbulence.

$$\mathbf{v} = \mathbf{v}_m + \mathbf{v}' \tag{3.14}$$

$$\phi = \phi_m + \phi' \tag{3.15}$$

$$\omega = \omega' \tag{3.16}$$

with subscript m describing the time-mean and a prime describing its perturbation. Note that the vertical velocity is zero on average (ω_m=0). This results in the total momentum equation

$$\frac{\partial \mathbf{v}'}{\partial t} + \mathbf{v}_m \cdot \nabla_p \mathbf{v}_m + \mathbf{v}' \cdot \nabla_p \mathbf{v}' + \mathbf{v}_m \cdot \nabla_p \mathbf{v}' + \mathbf{v}' \cdot \nabla_p \mathbf{v}_m + \omega' \frac{\partial \mathbf{v}_m}{\partial p} + \omega' \frac{\partial \mathbf{v}'}{\partial p} =$$

$$- \nabla_p \phi_m - \nabla_p \phi' - f\mathbf{k} \times \mathbf{v}_m - f\mathbf{k} \times \mathbf{v}' + \mathbf{F}_{\mathbf{r}m} + \mathbf{F}_{\mathbf{r}}'$$

$$\tag{3.17}$$

As the mean flow is defined to be steady, the local tendency of the mean velocity ($\partial \mathbf{v_m}/\partial t$) vanishes. In this study, the time-mean is defined as a 30-day average, centred on the investigation day (+/- 15 days from the day for which the analysis is performed, Harr and Dea, 2009). Averaging of the total momentum equation (Equation 3.17) over time leads to the time-mean momentum equation

$$\mathbf{v}_m \cdot \nabla_p \mathbf{v}_m + \overline{\mathbf{v}' \cdot \nabla_p \mathbf{v}'} + \overline{\omega' \frac{\partial \mathbf{v}'}{\partial p}} = -f\mathbf{k} \times \mathbf{v}_m - \nabla_p \phi_m + \mathbf{F}_{\mathbf{r}m} + \mathbf{F_0} \quad (3.18)$$

where $\mathbf{F}_{\mathbf{r}m}$ stands for the average of friction by time and $\mathbf{F_0}$ is introduced as a forcing term to compensate the insufficient steadiness of the time-mean, based on an only 30-day average. The perturbation momentum equation can now be obtained by subtracting the time-mean (Equation 3.18) from the total momentum equation (Equation 3.17). Merging the horizontal and vertical advection terms in the resulting eddy momentum equation (introducing the 3-d perturbation velocity $\mathbf{v}'_3$) and subsequently multiplying the whole equation by $\mathbf{v}'$ then finally leads to the desired budget equation for the eddy kinetic energy $K_e = (u'^2 + v'^2)/2$.

$$\frac{\partial K_e}{\partial t} + \mathbf{v}_m \cdot \nabla_p K_e + \mathbf{v}' \cdot \nabla_p K_e + \omega' \frac{\partial K_e}{\partial p} =$$
$$- \left(\mathbf{v}' \cdot \nabla_p \phi' \right) - \left(\mathbf{v}' \cdot \left(\mathbf{v}'_3 \cdot \nabla_3 \mathbf{v}_m \right) \right) \quad + \left(\mathbf{v}' \cdot \overline{\left(\mathbf{v}'_3 \cdot \nabla_3 \mathbf{v}' \right)} \right) - diss_e - \mathbf{v}' \cdot \mathbf{F_0}$$
$$(3.19)$$

Therein, the first term on the left hand side is the local tendency of the eddy kinetic energy, the second and third term are its advection with the mean and the eddy part of the total wind, respectively. The first term on the right hand side describes the advection of eddy geopotential heights by the eddy wind and is also referred to as pressure work term since in a Cartesian coordinate system the geopotential height in the pressure system equals pressure (Orlanski and Katzfey, 1991). It reveals the generation of (eddy) kinetic energy from (eddy) available potential energy due to the work done on an air mass by the horizontal pressure gradient force if there exist a cross-isobaric or ageostrophic acceleration component of the flow from regions with high towards regions with low geopotential (Kung, 1966, Kung and Baker, 1975). It should be noted that it is only the ageostrophic component of the eddy wind leading to generation of eddy kinetic energy. We will come back to this pressure work term later. The second and third terms on the right hand side are the energy conversions between the eddy and the mean flow, referred to as barotropic conversion or shear generation term (McLay and Martin, 2002), and the transfer between the eddy flow and the first-order correlation, which is zero if averaged by time (Orlanski and Katzfey, 1991). Frictional forces, $\mathbf{v} \cdot \mathbf{F_r}$ are abbreviated by *diss*, i.e. it is a dissipation term (4th term on rhs). The product of $\mathbf{v'} \cdot \mathbf{F_0}$, zero if averaged by time, can shown to be negligible on shorter time scales as well, as its maximum magnitude is about the magnitude of the velocity field (Orlanski and Katzfey, 1991). Assuming the mean wind to be non-divergent ($\nabla \cdot \mathbf{v}_m = 0$) and considering the continuity equation ($\nabla_p \cdot \mathbf{v'} = -\partial \omega' / \partial p$), the advection terms on the left hand side of Equation 3.19 can be simplified and summarised to

$$
\left.
\begin{aligned}
\mathbf{v}_m \cdot \nabla_p K_e &= \nabla_p \cdot (\mathbf{v}_m K_e) - K_e (\nabla_p \cdot \mathbf{v}_m) \\
\mathbf{v'} \cdot \nabla_p K_e &= \nabla_p \cdot (\mathbf{v'} K_e) - K_e (\nabla_p \cdot \mathbf{v'})
\end{aligned}
\right\} + \rightarrow \nabla_p \cdot (\mathbf{v} K_e) + K_e \frac{\partial \omega'}{\partial p}
$$

$$(3.20)$$

with $\mathbf{v}$ being the horizontal component of the total wind. Inserting this equation in Equation 3.19 leads to the final form of the eddy kinetic energy equation (cf. Orlanski and Sheldon, 1995)

$$
\frac{\partial K_e}{\partial t} = -\left(\mathbf{v'} \cdot \nabla_p \phi'\right) - \nabla_p \cdot (\mathbf{v} K_e) - \frac{\partial(\omega' K_e)}{\partial p} - \left(\mathbf{v'} \cdot \left(\mathbf{v'}_3 \cdot \nabla_3 \mathbf{v}_m\right)\right) + residue
$$

$$(3.21)$$

where the dissipation due to frictional forces, the first order correlation and the forcing term are casted in the residue.

The generation of eddy kinetic energy due to the pressure force (first term on right hand side of Equation 3.21) can be partitioned by separating the geopotential flux by the perturbation wind into a geostrophic and an ageostrophic geopotential eddy flux (Orlanski and Katzfey, 1991, Orlanski and Sheldon, 1995)

$$\mathbf{v}'\phi' = (\mathbf{v}'_a + \underbrace{\frac{\mathbf{k}}{f} \times \nabla_p \phi'}) \cdot \phi' \qquad \text{(3.22)}$$
$$\underbrace{\phantom{\frac{\mathbf{k}}{f} \times \nabla_p \phi'}}_{\mathbf{v}'_g}$$

In doing so the first term on the r.h.s. of Equation 3.21 can be reformulated as

$$-\mathbf{v}' \cdot \nabla_p \phi' = -\nabla_p \cdot \left(\mathbf{v}'\phi'\right) + \phi' \nabla_p \cdot \mathbf{v}' \qquad \text{(3.23)}$$

As the geostrophic flow is non-divergent, the divergence of the total geopotential eddy flux equals the divergence of the ageostrophic geopotential eddy flux. From the continuity equation it then follows that the div- or convergent ageostrophic flow is directly linked to vertical motions. Thus, using the hydrostatic equation ($\partial \phi' / \partial p = -\alpha'$) Equation 3.23 can be modified further

$$\begin{aligned}
-\mathbf{v}' \cdot \nabla_p \phi' &= -\nabla_p \cdot \left(\mathbf{v}'_a \phi'\right) + \phi' \nabla_p \cdot \mathbf{v}'_a \\
&= -\nabla_p \cdot \left(\mathbf{v}'_a \phi'\right) - \phi' \frac{\partial \omega'}{\partial p} \\
&= -\nabla_p \cdot \left(\mathbf{v}'_a \phi'\right) + \omega' \frac{\partial \phi'}{\partial p} - \frac{\partial \left(\omega' \phi'\right)}{\partial p}
\end{aligned}$$

$$\text{(3.24)}$$

Hence, the generation (destruction) of eddy kinetic energy results from the horizontal and vertical convergence (divergence) of the eddy ageostrophic geopotential flux (term 1 and 3 on the right hand side of Equation 3.24), which describe the dispersion of K_e by pressure work, and the baroclinic conversion of eddy available potential into eddy kinetic energy (eddy kinetic into eddy available potential) due to vertical motions (term 2 on rhs).

Consequently, the convergence of the ageostrophic geopotential flux can be interpreted as the horizontal import of potential energy into the region under consideration (Kung, 1977). It may be regarded as redistribution of the eddy kinetic energy, released by baroclinic conversion, into the regions of eventual generation (Kung, 1967). As will become clearer later, the ageostrophic geopotential flux always points downstream, due to mutual signs of the eddy ageostrophic flow and eddy geopotential disturbances in a wave pattern. Vertical motions generate kinetic energy if cold air (negative α') descents and warm air (positive α') rises.

In addition to investigations at individual pressure levels, the eddy kinetic energy budget and its contributing terms can provide valuable insight into the flow and the energy budget of the whole troposphere. Therefore, the eddy kinetic energy budget is vertically integrated over all available pressure levels. This leads to a more consistent picture of the horizontal energy distribution, its sources and sinks during the observed development, while terms describing the vertical redistribution of kinetic energy within an atmospheric column (the pressure deviation terms, i.e. term 3 on the rhs of Equation 3.24) become very small. Hence, as described in Harr and Dea (2009), a vertical integration is performed between the upper and lower pressure levels, respectively, p_t and p_s (typically the 100 hPa and 1000 hPa pressure levels). The resulting terms are then normalised by the gravitational acceleration $g = 9.81 m/s^2$ to get energy resembling units like joules per square-meter (J/m^2) for the eddy kinetic energy and watts per square meter (W/m^2) for its budget terms. Hence, the final equation to be used in the remainder of the study is as follows

$$\frac{1}{g}\int_{p_s}^{p_t}\frac{\partial K_e}{\partial t}\,dp =$$

$$\frac{1}{g}\int_{p_s}^{p_t}\left[-\left(\mathbf{v}'\cdot\nabla_p\phi'\right)-\nabla\cdot\left(\mathbf{v}K_e\right)-\left(\mathbf{v}'\cdot\left(\mathbf{v}'_3\cdot\nabla_3\mathbf{v}_m\right)\right)+residue\right]dp$$

$$\text{with}\quad\frac{1}{g}\int_{p_s}^{p_t}-\left(\mathbf{v}'\cdot\nabla_p\phi'\right)dp=\frac{1}{g}\int_{p_s}^{p_t}-\nabla_p\cdot\left(\mathbf{v}'_a\phi'\right)dp-\frac{1}{g}\int_{p_s}^{p_t}\omega'\alpha'\,dp$$

$$(3.25)$$

To summarise the eddy kinetic energy budget equations and to ease refer-
encing when discussing the results from applying the budget on several data
sets in the remainder of this work, all relevant terms, their names and abbreviations are listed in Table 3.1. With the abbreviations defined in Table
3.1, Equation 3.25 can be written in another way

$$K_e\text{-Budget:}\qquad \frac{\partial K_e}{\partial t}\quad = \quad -generation-div\text{-}keflux-brtr+residue$$

$$\text{with:}\quad generation \quad = \quad div\text{-}ageo+bcli$$

b) Atmospheric Developments Seen in an Eddy Kinetic Energy Framework

Various studies conducted the analysis of the eddy kinetic energy budget
to investigate sources and sinks of kinetic energy in distinct atmospheric
developments. Orlanski and Katzfey (1991) could show that eddy kinetic
energy dispersion via energy fluxes from upstream regions provided a major source of eddy kinetic energy for the development of a cyclone wave
in the Southern Hemisphere. Furthermore, they could identify a sequential
propagation of eddy kinetic energy centres into downstream regions. On

Table 3.1: Different terms of the eddy kinetic energy budgets (left, from Equation 3.25), their abbreviations as used in the remainder of the thesis (middle) and their meaning (right). As in Equation 3.25, all terms are vertically integrated, but integral terms are not shown to ease reading of the table.

Term	Abbreviation	Description
$\frac{1}{2}(u'^2 + v'^2)$	K_e	eddy kinetic energy
$\frac{\partial K_e}{\partial t}$	-	local tendency of K_e
$\mathbf{v}' \cdot \nabla_p \phi'$	*generation*	total generation of K_e
$\omega' \alpha'$	*bcli*	baroclinic conversion
$\mathbf{v}' \cdot (\mathbf{v}'_3 \cdot \nabla_3 \mathbf{v}_m)$	*brtr*	barotropic conversion
$\mathbf{v}'_a \phi'$	*ageo*	ageostrophic geopotential eddy flux
$\nabla_p \cdot (\mathbf{v}'_a \phi')$	*div-ageo*	divergence of the *ageo*
$\mathbf{v} K_e$	*keflux*	K_e flux by the total wind
$\nabla_p \cdot (\mathbf{v} K_e)$	*div-keflux*	divergence of *keflux*
$\mathbf{v} K_e + \mathbf{v}'_a \phi'$	*totflux*	total flux of K_e

the other hand, the energy fluxes also played a crucial role in dissipating the system. Using a simple primitive equation model, Orlanski and Chang (1993) identified the ageostrophic geopotential flux convergence as the predominating propagator of eddy kinetic energy in downstream as well as upstream development in baroclinic waves. These findings were underscored by Chang and Orlanski (1993), who investigated the dynamics, maintenance and propagation of a storm track (Hoskins and Valdes, 1990) over the North Pacific basin in a simple primitive equation model and supported their findings with ECMWF analysis data. Warm ocean currents just off the Japanese coast caused increased baroclinicity and thus baroclinic development of cyclonic systems in the western North Pacific. These developments were accompanied by increasing kinetic energy and an amplification of the wave pattern. In the mature stage of the baroclinic development, downstream radiation by the ageostrophic geopotential flux propagated eddy kinetic energy into downstream regions, which are typically characterised by weaker baroclinicity. Dispersed eddy kinetic energy from the upstream region then triggered and maintained the development of new baroclinic systems or eddies in this low-baroclinicity downstream environment. These results indicate energy fluxes, like the ageostrophic geopotential flux to apparently be essential for the propagation and maintenance of storm-tracks in regions, which normally would have no sources for baroclinic development. Using eight years of winter data, Chang (1993) could further prove the importance of the ageostrophic geopotential flux as he identified downstream development processes in the Northern Pacific storm track to dominate local baroclinic conversion. These findings were later corroborated by Danielson et al. (2004), who investigated forty-one cold season cyclones in the eastern North Pacific. About half of the cyclones investigated were clearly triggered by eddy kinetic energy originating from cyclonic systems in the upstream side of the North Pacific basin, which already existed some hours or days before the observed development in the eastern North Pacific. In contrast, those upstream cyclones in the western North Pacific often developed due to local baroclinic processes (i.e. rising of warm air) instead of energy fluxes from upstream regions. At the same time, energy dispersion towards the downstream eastern cyclones constituted an important factor for cyclone decay in the western North Pacific.

With the outcome of those studies in mind, Orlanski and Sheldon (1993) performed a case study to prove the concept of downstream propagating eddy activity, which they entitled "downstream baroclinic development".

They simulated the development of a strong cyclone associated with a amplified wave pattern in the North Pacific storm track. In this case, baroclinic processes supported the development and intensification of the cyclone in downstream regions only at later and mature stages. The initial activation for the cyclone to form was by converging energy fluxes from upstream regions. Based on all these preliminary studies and their indication on how to explain downstream development in an energy frame work, Orlanski and Sheldon (1995) finally developed their conceptual paradigm about "downstream baroclinic development".

In the remainder of this Chapter, "downstream baroclinic development" and the involved energy conversions are explained in a theoretical frame work, followed by an application to a real data set to analyse downstream propagation of Rossby wave trains. Later, this approach will be used to diagnose how a transitioning tropical cyclone influences the mid-latitude flow pattern (Chapter 5).

3.2.3 The Downstream Baroclinic Development Paradigm

From the set of equations derived above, it becomes apparent that multiple quantities are influencing the budget of eddy kinetic energy and thus may play a role in downstream baroclinic development. Closer examination of the individual quantities now provides insight into the physical processes affecting downstream development.

a) The Ageostrophic Geopotential Flux and K_e Generation

The basic idea of downstream baroclinic development and especially of the ageostrophic geopotential flux can be explained by investigating a fully developed trough, according to Orlanski and Sheldon (1995), whose schematic illustration is reproduced as Figure 3.2. This trough is characterised by an already decaying maximum of eddy kinetic energy (a so-called K_e centre) at its western flank and a growing eddy kinetic energy centre at its eastern flank. In this notation, centre stands for a region with enhanced K_e in general, while maximum denotes the K_e centre of the wave train which is strongest at the time of investigation. The relative strength of the centres is indicated by the number of ellipses. The K_e centres always coincide with the flanks of the wave pattern, where strongest winds are observed. They often resemble the so-called jet-streaks (Orlanski and Sheldon, 1995).

Compared to the mean flow, which is typically more zonal than the wave pattern, we have negative geopotential deviations in the trough ($\phi' < 0$) and positive deviations in the ridge ($\phi' > 0$). Due to the centrifugal force, the wind at upper levels (black arrows, Figure 3.2) will be super-geostrophic in the crest of the ridges and sub-geostrophic in the base of the trough (Lim and Wallace, 1991), as indicated by the open arrows in Figure 3.2. This causes a negative ageostrophic wind ($\mathbf{v_a} < 0$) in the troughs and positive ($\mathbf{v_a} > 0$) in the ridges (open arrows, Figure 3.2). Hence, the ageostrophic geopotential flux (*ageo*), which is defined as a product of those two quantities, is always positive and points into the downstream direction. The flux is maximised in the centres of the troughs and ridges and points from the decaying maximum of K_e on the rear ridge flank towards the growing K_e centre on the front flank of the ridge. The exit regions of the energy centres are thus typically characterised by divergence of the ageostrophic geopotential flux (Figure 3.3), which can be seen to remove K_e from this centre. In turn, flux convergence leads to an increase of kinetic energy in the entrance regions of the K_e centre (Figure 3.3).

By discussing the K_e centres as jet-streaks, the processes of the K_e generation term may become more perspicuous. In the entrance region of a jet-streak, the flow experiences an increased pressure gradient force, which can not be compensated by the Coriolis force at the onset of the acceleration. This initially causes an ageostrophic wind component from higher to lower pressure, constructive work is done by pressure forces. Divergence in the right entrance region and convergence in the left entrance region induce a direct vertical circulation that converts eddy available potential into eddy kinetic energy. The flow is further accelerated into the jet maximum. In the exit region, the pressure gradient force is reduced, but at the onset of deceleration, the Coriolis force exceed the pressure gradient force and induces an ageostrophic flow component from lower to higher pressure. Thus, destructive work is done by the pressure gradient forces and the flow is decelerated. The divergence and convergence pattern now induce a thermal indirect circulation which converts eddy kinetic into eddy available potential energy in the exit region (after Persson, 1998). The ageostrophic flow from low to high pressure in the exit regions further enhances the subsequent trough or ridge and can cause the development of another jet-streak further downstream, where air flows again from high to low pressure and eddy available potential energy is converted back into eddy kinetic energy.

b) The Flux of K_e with the Total Wind

This term describes the flux of eddy kinetic energy itself with the total wind ($keflux$). It is strongest within a K_e maximum, with divergence in the entrance and convergence in the exit region (Figure 3.4) and redistributes K_e within the K_e centre. The flux vector can be separated into a zonal and a meridional component. The flux in the zonal direction causes the whole K_e centre to move eastwards with the phase velocity of the wave, while the meridional component redistributes K_e from the entrance to the exit region of the K_e centre, from where it is in fact instantaneously dispersed by the divergent ageostrophic geopotential flux towards the downstream centre (Figure 3.3)

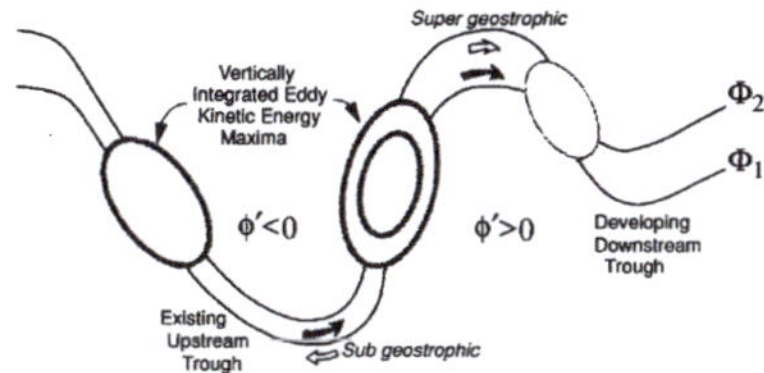

Figure 3.2: Different components of a baroclinic wave. The upper-level geopotential height field is indicated by contours, centres of enhanced vertically averaged K_e are marked as ellipses. Number of ellipses defines strength of the centre. Black arrows depict air flow relative to the wave, the ageostrophic wind is indicated by the open arrows. Eddy geopotential height is positive in the ridge and negative in the trough. Figure 2 from Orlanski and Sheldon (1995).

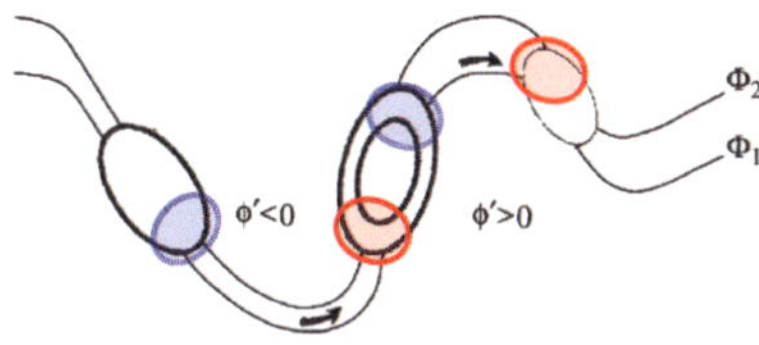

Figure 3.3: Schematic illustration for the ageostrophic geopotential flux and its divergence and convergence. Flow structure and K_e centres as in Figure 3.2. Convergence of *div-ageo* in red, divergence in blue.

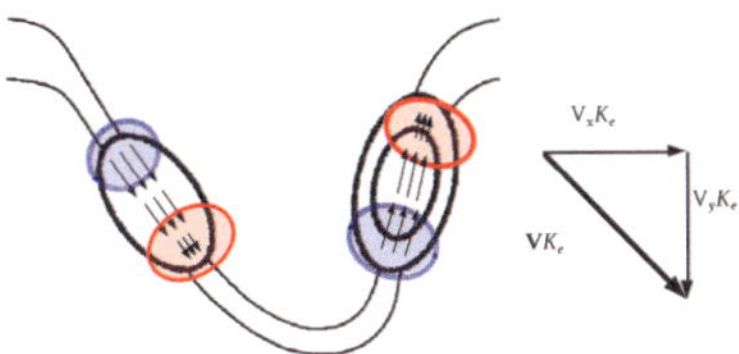

Figure 3.4: Schematic illustration for the flux of K_e with the total wind (arrows) and its divergence and convergence. Flow structure and K_e centres as in Figure 3.2. Convergence of *div-keflux* in red, divergence in blue. The flux component can be separated into a zonal (x) and meridional (y) component, highlighting the propagation of the K_e centre.

c) Downstream Baroclinic Development

The individual terms can now be brought together to explain the full cycle of downstream baroclinic development (Figure 3.5). According to Orlanski and Sheldon (1995), the process of downstream baroclinic development starts with a pre-existing K_e maximum. This mature K_e centre might have developed due to enhanced baroclinic processes (ascending warm and descending cold air masses) at a frontal zone, as discussed by Chang and Orlanski (1993), and possibly (as in our case) due to converging ageostrophic geopotential fluxes from upstream regions as well. Ageostrophic geopotential fluxes (large open arrows in Figure 3.5) now emanate from this mature centre (the centre already loses some K_e, as indicated by the - sign) and disperse eddy kinetic energy through the ridge towards the downstream centre W. This dispersion of K_e results from an increase in the geopotential height gradient and a further amplification of the ridge, as eddy kinetic is converted into eddy potential energy due to ageostrophic flow from lower towards higher heights. Convergence of the ageostrophic geopotential flux in the entrance region of centre W coincide with a flow component from higher towards lower pressure, and the pressure gradient force further accelerates the flow through the growing K_e centre W (step 1). Eddy available potential energy is converted into eddy kinetic energy and the new K_e centre W grows further. Furthermore, convergence of the upper level ageostrophic flow (blue arrows in Figure 3.5 point towards each other) on the rear flank of this incipient trough induces subsidence in the cold anomaly and converts eddy available potential into eddy kinetic energy (sinking cold air masses). Thus, after the development of the new downstream centre W is initialised by the ageostrophic geopotential flux convergence, this centre experiences further growth due to baroclinic conversion. The acceleration of the flow through centre W supports the deepening of the trough and the ageostrophic wind in the base of the trough increases. An ageostrophic geopotential flux establishes in the base of the trough, starts to remove eddy kinetic energy

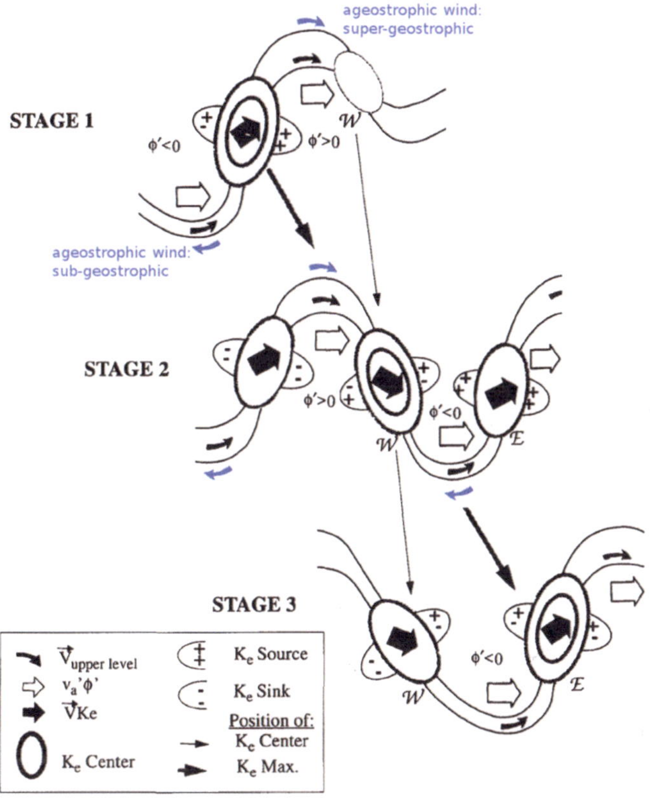

Figure 3.5: Schematic illustration of downstream baroclinic development (with slight modifications reproduced from Orlanski and Sheldon, 1995). In the three stage process the upstream maximum (ellipse) decays, while the downstream centres W and E gain K_e (+/- signs) due to convergence of the ageostrophic geopotential flux (open arrow). The ageostrophic geopotential flux always points downstream due to sub- or super-geostrophic total winds (blue arrows) and positive or negative geopotential anomalies in the troughs and ridges, respectively. The K_e maximum propagate with group velocity (long thick arrow), a single K_e centre, on the other hand, propagates with phase velocity (long thin arrow)

from the exit region of this actual K_e maximum W (- signs indicate onset of loss) and propagates it through the trough towards another newly developing centre E on the front flank of this trough (stage 2). Ageostrophic flow from lower to higher heights in the exit region of centre W decelerates the flow and enhance the geopotential gradient in the base of the trough. In the entrance region of centre E ageostrophic flow points towards lower height and enables conversion from eddy available potential into eddy kinetic energy. Ascent of warm air on the eastern trough flank (due to divergent ageostrophic wind in upper levels) supports the further growth of centre E, as is also happened in centre W. Centre E has not started to disperse K_e further downstream in stage 2, hence it only experiences net generation (only + signs in centre). For both centres, the baroclinic processes start well after the initialisation of the energy centre due to the convergence of the ageostrophic geopotential flux as it was shown by Chang and Orlanski (1993) and Orlanski and Sheldon (1993).

As the predominant energy source of centre W (energy fluxes from the upstream centre) cease to exist, centre W will also start to decay gradually, due to dispersion of energy towards the downstream regions. However, kinetic energy generation due to baroclinic conversion decelerates the decay of centre W for some time. The downstream development will continue until the energy sinks (ageostrophic geopotential fluxes towards downstream regions and Reynolds stresses which are here not discussed further) overwhelm the energy accumulation due to baroclinic conversion and convergence of energy fluxes from upstream regions. During the whole cycle of downstream baroclinic development, the meridional component of flux of K_e with the total wind $keflux$ redistributes K_e from the rear to the front of the K_e centres and hence support the downstream dispersion of K_e. At the same time, the zonal $keflux$-component causes the whole energy centre to move eastwards with the phase speed of the wave (thin long arrows in Figure 3.5). This instantaneous dispersion of kinetic energy through the trough provides an explanation for the downstream propagation of the mature kinetic energy maximum by group velocity (thick long arrows in Figure 3.5).

As highlighted above, analysis of the eddy kinetic energy budget and its constituent components provides useful insight into the downstream propagation of wave disturbances. It facilitates an explanation for how baroclinic development may be triggered in regions with weaker baroclinicity. In applications, this analysis technique now enables the identification of kinetic energy sources for baroclinic developments and allows to investigate the role of neighbouring systems. Based on case studies of four tropical cyclones, Harr and Dea (2009) first identified the role of a transitioning tropical cyclone in the this framework. A tropical cyclone, typically characterised by strong winds and enhanced baroclinic processes, forms an additional source of eddy kinetic energy. If a tropical cyclone undergoes extratropical transition and interacts with the mid-latitude flow, ageostrophic geopotential fluxes might disperse eddy kinetic energy from the cyclone into the mid-latitude flow. As baroclinic conversion of available potential into eddy kinetic energy might be strong during the transitioning process[4], the baroclinic conversion help to maintain the strong K_e signal of the extropical cyclone. In this way, a preexisting energy centre will gain additional K_e dispersed from the transitioning storm or even a new K_e centre will be triggered. The additional K_e can then furthermore benefit downstream development. Hence, the ET of a TC might support downstream baroclinic development or may even initiate it.

Motivated by the study of Harr and Dea (2009), the K_e budget is in this work applied to several possible development scenarios that resulted from a clustering of the ECMWF EPS. Thereby, the objective is to identify and further examine the underlying physical and dynamical processes that cause the differences in downstream baroclinic development in the ensemble fore-

---

[4]Warm, humid subtropical air rises at the mid-latitude baroclinic zone north and east of the transitioning cyclone, due to cyclonic rotation of the storm (see e.g. Klein et al., 2000)

cast. This will provide further insight into the interaction between the transitioning storm and the mid-latitude flow.

3.2.4 Downstream Baroclinic Development and Rossby Wave Trains

Prior to this, a short excursion into the detection and propagation of large scale atmospheric wave packages, so-called Rossby wave trains, will allow to highlight and further explain the application of the K_e budget analysis method. Mid-latitude circulation patterns are often characterised by a series of adjacent pressure systems, which are organised in storm tracks (Hoskins and Valdes, 1990). The associated wave packages in the upper troposphere then form a so-called Rossby wave train. Such Rossby wave trains might be precursors for severe weather events like the Elbe flooding in 2002 (Grazzini and van der Grijn, 2003) or heavy precipitation south of the Alps (Martius et al., 2008). As they also often impact predictability for mid-latitude regions (Shapiro and Thorpe, 2004), Rossby wave trains are some of the objectives to be further investigated in the THORPEX program. Thus, the identification and further examination of Rossby wave trains and their linkage to predictability for the Atlantic-European sector also lies within the scope of the German research group PANDOWAE (Predictability and Dynamics of Weather systems in the Atlantic-European Sector), being a part of THORPEX. The different approaches undertaken in PANDOWAE to assess the predictability and identification issues are systematically compared and discussed in a joint publication about the *"Utility of Hovmöller diagrams to diagnose Rossby wave trains"* by Glatt et al. (2011). Therein, each of the employed analysis techniques results in Hovmöller diagrams (Hovmöller, 1949) of one or more quantities. These Hovmöller diagrams then enable the identification of wave packages, like Rossby wave trains, during the two-month sample period September-October 2008. As the individual budget terms of the EKE-analysis clearly provide insight into the development and downstream propagation of wave patterns, Hovmöller diagrams for the EKE-budget form one part of the comparison project. The main outcome of applying the EKE analysis on the sample period and of the comparison project in general will now be summarised briefly.

a) Sample Period and Data Base

The sample period for the joint case study was chosen to span September and October 2008. During this time, strong upper-level Rossby wave activity could be observed. Furthermore, several tropical cyclones (Hurricanes Gustav, Hanna, Ike and Typhoons Sinlaku, Jangmi and Bavi) underwent ET. Thus, in addition to the conventional downstream propagation of Rossby wave trains, the period also allows to examine the influence of transitioning tropical cyclones on the mid-latitude flow.

The data base for the joint case study was the operational ECMWF analysis in $0.5°$ horizontal resolution with 14 vertical pressure levels and a 12 hour time increment. For the whole sample period the EKE budget (as described in Equation 3.21 and Table 3.1) was computed against the background of a 30-day running mean state (derived from analysis data as well). Then, Hovmöller diagrams of the individual terms in Equation 3.21, averaged between $40°N$ and $60°N$, indicate the contribution of these terms to the formation and propagation of wave packages.

b) Identification of Rossby Wave Train Propagation

In general, Rossby wave trains represent a deviation of the actual mid-latitudinal flow pattern from the 30-day average (zonal mean flow) and thus coincide with enhanced K_e. The Hovmöller diagram (Figure 3.6a) highlights the location, intensity and downstream propagation of local K_e

centres. As discussed above (Section 3.2.3), these K_e centres typically coincide with the flanks of the troughs and ridges in a mid-latitude wave pattern. Thus, this approach allows to clearly identify subsequent wave patterns in the mid-latitudes. The overall increase in K_e towards the end of October arise due to the generally enhanced wave activity during the fall and winter season, compared to the more quiet summer season. As shown in Glatt et al. (2011), the region between $0°$ and $90°$W was prone to wave breaking events[5] during the whole period, which explains the lack of enhanced K_e.

For most of the strong K_e centres in Figure 3.6a, the net generation of K_e (*generation*, Figure 3.6b) is in part attributable to the baroclinic conversion (*bcli*, Figure 3.6c) of eddy available potential into eddy kinetic energy due to vertical motions. Although most regions with strong net *generation* coincide with regions of enhanced *bcli*, it becomes obvious that convergence of the ageostrophic geopotential flux (*div-ageo*, Figure 3.6d) acts as a strong source as well. This term e.g. explains the development of the strong K_e centre just west of $150°$W between 27 and 29 Sep 2008. Originally, baroclinic conversion is strongest east of $180°$W. However, the associated K_e centre there stays rather weak, as the ageostrophic geopotential flux already disperses (divergence, blue) K_e towards the downstream centre (convergence, red). The barotropic conversion (*brtr*, Figure 3.6f) and thus the transfer of K_e between the eddy and the mean flow is negative in most regions. Especially in the regions of enhanced wave breaking (Greenwich meridian), K_e is lost to the mean flow. The remaining quantity, the convergence and divergence of K_e flux with the total wind (*keflux*, Figure 3.6e) is typically aligned around a single K_e centre. Its role will become more clear in the next example. The ET of Hurricane Hanna (around 7 September near $75°$W, $40°$N) and Hurricane Ike (around 14 September near $90°$W, $35°$N) coincided with slightly enhanced K_e, *generation* and *bcli* signals,

---

[5]Wave breaking means the wave pattern wraps up cyclonically or anticyclonically. This wrap-up then causes the decay of the wave instead of a further downstream propagation.

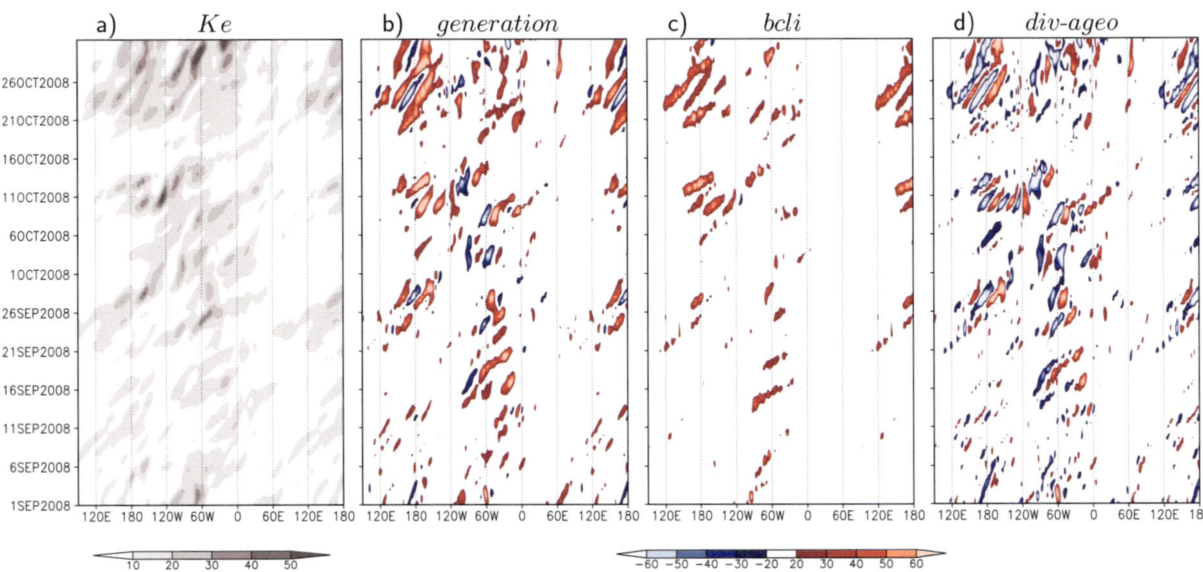

Figure 3.6: Hovmöller diagrams for sample period September-October 2008, averaged between 40°N and 60°N, for eddy kinetic energy and the budget terms listed in Table 3.1. Red colours express convergence, gain of K_e, blue colours indicate divergence, loss of K_e. a) eddy kinetic energy (K_e) in $10^5 J/m^2$, b) total generation of K_e (*generation*) in W/m^2, c) baroclinic conversion of K_e (*bcli*) in W/m^2, d) divergence of ageostrophic geopotential flux (*div-ageo*) in W/m^2. Area between 90° and 180°E is plotted twice to ease identification of connected signals. Adapted from Glatt et al. (2011). Continued on next page.

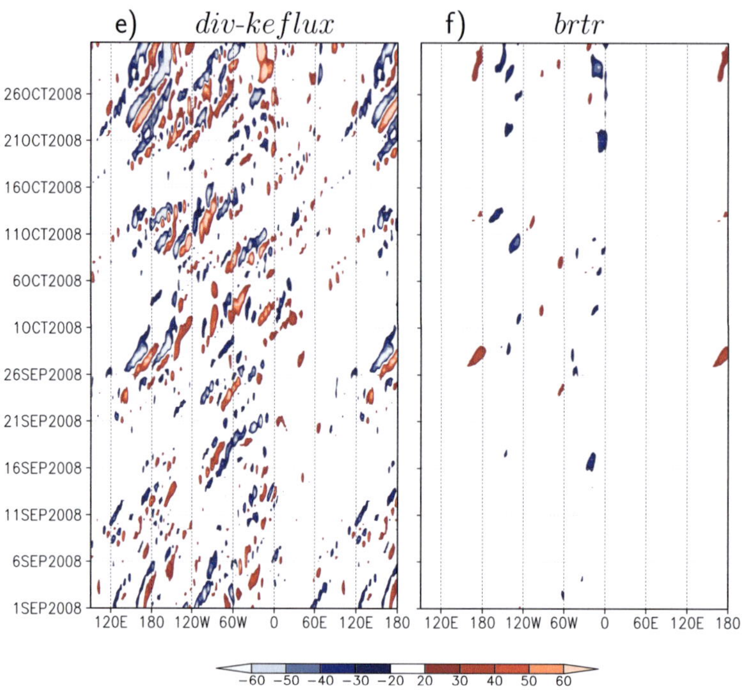

Figure 3.6: continued. e) convergence of K_e flux with the total wind (*div-keflux*) in W/m^2, f) barotropic conversion of K_e (*brtr*) in W/m^2

emanating from the ET regions, but their impact on amplification of the mid-latitude flow was not really strong. Typhoon Sinlaku and Typhoon Jangmi stayed rather south of 40°N (southern boundary of average belt). Hence, their albeit rather weak impact on the mid-latitude flow might be mis-represented in this approach. The ET of Tropical Storm Bavi (ET around 21 October near 160°E, 40°N) was followed by a clear increase in K_e, *bcli*, and *generation* (Figure 3.6a-c, 20-24 September, around 180°E) which was partly attributable to the transitioning storm, but also to support from upstream regions. A closer examination of the individual terms in a finite temporal and spatial region provide further insight into the processes during downstream baroclinic development. Therefore, we focus on the three adjacent K_e centres (labelled in Figure 3.7a) that expand from $120°\text{W}$ to $300°\text{W}$ between the 21st and 28th of September. The first K_e centre (starting at $110°$) forms the K_e maxima of the observed wave pattern until approximately 12 UTC on 24th of Sep. In the mean time it has an only weak gain in K_e (total generation term, Figure 3.7b), mainly due to converging ageostrophic geopotential flux. There exists some baroclinic conversion of Ke (Figure 3.7c), but most of this additional K_e is already dispersed further downstream by the diverging ageostrophic geopotential flux (Figure 3.7d, blue contours). However, the total flux clearly acts to redistribute K_e inside this centre from its rear (divergence, blue) to its front flank (convergence, Figure 3.7e). By expanding the examination onto the subsequent K_e centres (2 and 3), the downstream propagation becomes obvious. Apparently, centre 2 initially grows through convergence of the ageostrophic geopotential flux at its western edge (Figure 3.7d, $80°\text{W}$ around 24 Sep). The additional K_e is then redistributed by the *keflux* from the rear to the front of the maximum (Figure 3.7e). The centre becomes K_e maximum around 12 UTC on the 24th of September.

The baroclinic conversion of K_e in the new maximum do not grow until this centre is well established (Figure 3.7 c) by the ageostrophic geopotential flux (later on 24th of Sep). On the other hand, the ageostrophic geopotential fluxes quite soon disperse K_e from centre 2 further downstream into centre 3, where strong flux convergence (Figure 3.7 d) causes a net generation of K_e (Figure 3.7 b). Through the whole period, the characteristic contribution of K_e transport by the total wind, namely the redistribution of K_e inside a centre from its rear to its front, stands out clearly. Finally, barotropic conversion from the eddy to the mean flow comes into play at the very end of the development under investigation, when it causes the a sink for K_e in centre 3.

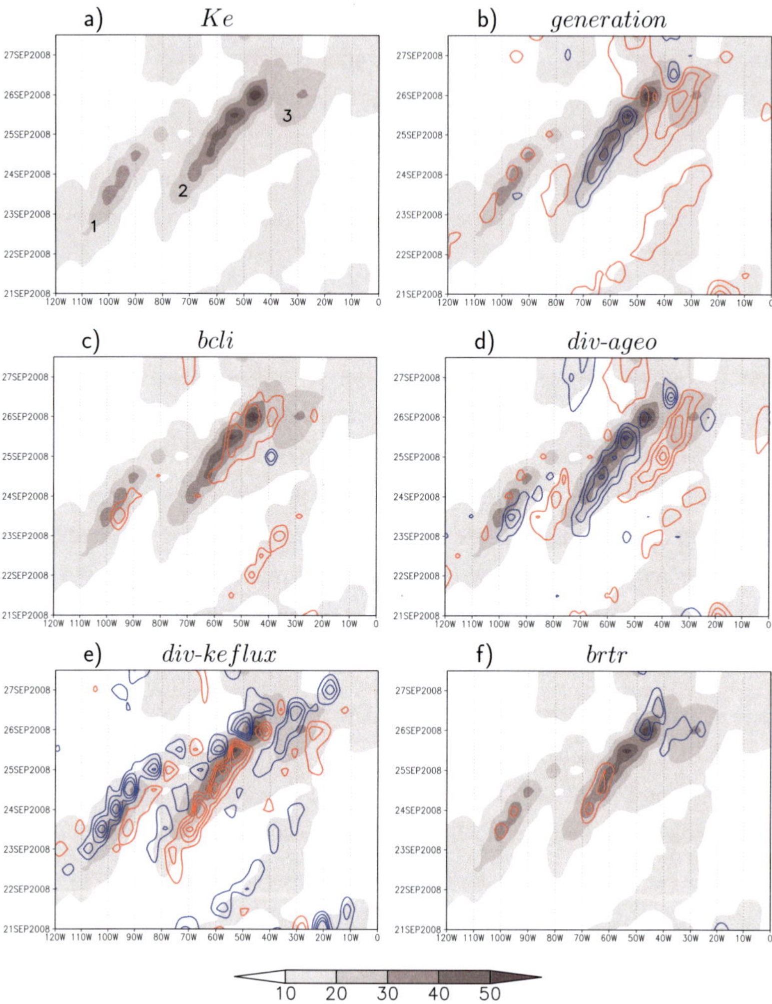

Figure 3.7: Sections of Hovmöller diagram for a specific K_e signal. Ke in shading, different source and sink terms in contours (red=gain, convergence, blue=loss, divergence): a) eddy kinetic energy K_e, b) total generation of Ke (*generation*), c) baroclinic conversion (*bcli*), d) convergence (positive) of the ageostrophic geopotential flux (*div-ageo*), e) convergence (positive) of K_e flux with total wind (*div-keflux*) and f) barotropic conversion (*brtr*). K_e in $10^5 J/m^2$, budget terms in W/m^2.

c) Identifying Rossby Wave Trains in Hovmöller Diagrams

The discussion of this sequence highlighted the basic idea and the involved processes of downstream baroclinic development using real analysis data. However, the question still to answer is the achievement of the EKE analysis approach in the comparison project for identifying Rossby wave trains. Three kinds of Hovmöller diagrams which allow the identification of Rossby wave trains were compared systematically in Glatt et al. (2011). Thereby, the focus was put on the distinct representation of Rossby wave trains in the individual diagnostic tools. Furthermore it was considered, which of the the distinct diagnostic tools provide useful insight into the development and propagation of Rossby wave trains at different stages of their life cycles. Overall, the study aimed to develop objective identification methods to operationally identify Rossby wave trains in large data sets. This would enable climatological studies about the distribution and occurrence of Rossby wave trains and further may point to their linkage to high impact weather and prediction issues.

For the sample period (September-October 2008), Rossby wave trains were identified in several types of Hovmöller diagrams: conventional, but adapted trough- and ridge Hovmöller diagrams, which provide full phase information, Hovmöller diagrams which neglect phase information and treat the wave trains as whole individual objects, and in Hovmöller diagrams that furnish insight into physical processes during the life cycle of Rossby wave trains. The EKE analysis pertains to the third group of Hovmöller diagrams. During investigation, each method turned out to have a characteristic behaviour, advantages and disadvantages. The full-phase Hovmöller diagrams, a conventional trough-ridge diagram, a latitudinally weighted trough-ridge diagram and a trough-ridge diagram which follows the wave guide, provided good means to identify most of the Rossby wave trains during their propagation. Diagrams, which consider a wave train as an entity (identifying the envelope of wave trains or the zonal component of the Plumb flux (Plumb, 1985), were sensitive to selection of parameters and length of the waves to investigate. They often enabled identification of Rossby wave trains found with other methods, but they also identified additional wave train objects that were not depicted by other methods. Hovmöller diagrams of different terms from the EKE analysis provided useful insight into development and propagation of Rossby wave trains. They furthermore indicate whether local wind maxima belong to the same wave train or not. Finally, two distinct diagnostics highlight different phases of Rossby wave breaking events, which typically occur at the end of the life cycle.

Although the study could highlight and characterise advantages and disadvantages, it was not possible to assign only one specific method to identify objectively Rossby wave trains from Hovmöller diagrams. Hence, selection of an adequate method to identify Rossby wave trains depends on the features of Rossby wave trains which shall be investigated.

This application of the EKE analysis demonstrates the ability of the method to identify downstream propagation of wave patterns and to highlight the processes involved. Therefore, it shows promise to provide useful insight into differences in downstream development which cause the distinct synoptic scenarios found in the TIGGE or an individual ensemble system using the EOF- and Cluster analysis.

4 Characteristics of the TIGGE Multimodel Ensemble in Representing Forecast Variability Associated with Extratropical Transition

Using the variability among ensemble members as a proxy for predictability during ET events, Anwender (2007), Anwender et al. (2008) and Harr et al. (2008) could show that the forecast uncertainty increased after the completion of an ET event and propagated towards downstream regions. By applying the EOF- and clustering approach, described in Section 3.1.1 to ensemble forecasts from the ECMWF and the NCEP EPS they identified typical spacial patterns of variability related to the ET of a number of tropical cyclones.

Although these stand-alone ensemble forecasts already consider distinct error sources like uncertainties in the initialisation or even in the parametrisations, they still may suffer from errors that arise due to approximations in the model formulation, the underlying physics or the incorrect representation of initial uncertainties. One way to overcome biases due to the different model approaches is provided by the use of multimodel ensemble forecasts, like the TIGGE data set, which was introduced in Section 2.2.2 (cf. Table 2.1). This new data set offers the possibility to extend the studies by Anwender (2007), Anwender et al. (2008) and Harr et al. (2008) by considering a variety of different ensemble systems. In the first part of the work at hand, the analysis method from Harr et al. (2008) is adapted to the TIGGE multimodel ensemble for the first time to identify forecast uncertainty related to ET and to characterise the behaviour of the multimodel ensemble in forecasting ET events. Specific questions to be answered during the study are as follows. Does the combination of several EPS in TIGGE provide an increased representation of forecast variability? Through this increased variability can a broader range of possible future states of the atmosphere be identified by using the TIGGE multimodel ensemble instead of using only a single ensemble system? Directly connected to this point is the question, as to how the individual ensembles contribute to this variability, i.e. whether there are some EPS which span the whole phase space of the multimodel forecast while others mainly focus on one specific part of it? Some of the ensembles use the same perturbation methods to create the ensemble, e.g. Australia, Japan and ECMWF use singular vectors, while other centres perturb their ensembles differently. Hence, another important point to address is whether some of the constituent EPS often show related development scenarios which will lead to preferred groupings during the cluster analysis and if this can be linked to the perturbation method used. The outcome of this study can also be found in Keller et al. (2011).

4.1 Data Base of Study

To answer these questions, three different data sets are analysed: TIGGE as the whole multimodel ensemble, the ECMWF EPS as an individual EPS, and the TIGGE multimodel without the ECMWF EPS, referred to as TI-EC in the remainder. Thereby, eight of the ten TIGGE EPS are considered (Table 2.1). The MeteoFrance- and the Korean Meteorological Agency (KMA) EPS are not included, due to the short forecast time of the former and a systematic bias in the latter. Evidence for the problem with the KMA EPS arose in the early stages of this work. Members from the KMA EPS often formed outliers in the cluster analysis and showed distinct development scenarios from all of the other members. Closer examination of the fields under investigation showed a comparatively low geopotential height at 200 hPa in the southern mid-latitudes and subtropics. Personal communication with the contact person of KMA responsible for their TIGGE contribution confirmed these findings as being consistent with a systematic bias in their global model. The TIGGE data set under investigation includes the EPS from Australia, Brazil, China, Canada, ECMWF, Japan, NCEP and UK Met Office. This gives a multimodel EPS with 231 members, interpolated on a $1°x1°$ grid and with 12 hourly data output. The forecast variable under consideration in this study is the 500 hPa geopotential height, as it nicely captures the interaction between the transforming storm and the mid-latitude flow. The potential temperature at the dynamic tropopause, which was used by Harr et al. (2008), Anwender et al. (2008) and Harr et al. (2008) is only available for three of the constituent TIGGE EPS and thus could not be analysed for this study.

Using this database, ten forecasts are investigated that were initialised prior to the completion of ET of five tropical cyclones (two hurricanes and three typhoons) in 2008 (Table 4.1), the first year for which the full TIGGE multimodel ensemble was available. The tropical cyclones investigated and their characteristics during ET are introduced in Section 2.4. All cases have in common that they led to a decrease of predictability for downstream regions at least in a couple of ensemble forecasts during their ET process. The forecasts used are initialised at the latest two days before the system was declared as extratropical by the Regional Specialised Meteorological Centres (RMSC). All of them show an enhanced variability in the forecast and thus a loss of predictability in the region around and downstream of the transitioning storm. The choice of initialisation time was motivated by Harr et al. (2008), who showed the forecast uncertainty to decrease after ET. ECMWF is acknowledged as one of the leading centres in global ensemble forecasting. Furthermore, this EPS was already used by (Anwender et al., 2008) to investigate predictability and development scenarios during ET events. For these reasons this EPS was chosen as the single model EPS to examine in this study. In the remainder of this section, the method of

Table 4.1: Overview of variability in the five ET cases studied (H: Hurricane, TY: Typhoon, TS: Tropical Storm). (adapted from Keller et al., 2011).

			Forecast	
ID	Storm	Name	Initialisation Time	Investigation Time
1	H Hanna	Hanna05	05 Sep 08, 12Z	09 Sep 08, 00Z
2	TY Jangmi	Jangmi26	26 Sep 08, 12Z	02 Oct 08, 12Z
3	H Ike	Ike12	12 Sep 08, 12Z	15 Sep 08, 12Z
4	TS Bavi	Bavi17	17 Oct 08, 12Z	20 Oct 08, 00Z
5	TY Sinlaku	Sinlaku16	16 Sep 08, 12Z	22 Sep 08, 00Z
6	TS Bavi	Bavi18	18 Oct 08, 12Z	20 Oct 08, 00Z
7	H Ike	Ike11	11 Sep 08, 12Z	15 Sep 08, 12Z
8	TY Sinlaku	Sinlaku15	15 Sep 08, 12Z	21 Sep 08, 12Z
9	TY Sinlaku	Sinlaku14	14 Sep 08, 12Z	22 Sep 08, 00Z
10	H Ike	Ike10	10 Sep 08, 12Z	15 Sep 08, 12Z

analysis and interpretation of the results is demonstrated for one sample case, then the outcome for all ten cases is summarised and generalised to provide a coherent picture of the results. A 9-day ensemble forecast, which was initialised at 12 UTC on 10 September 2008 and thus 96 h prior to the ET of Hurricane Ike, referred to as Ike10, serves as sample case as it nicely highlights the different aspects of the outcome of the study.

4.2 Differences in Forecast Uncertainty

The uncertainty contained in an ensemble forecast can be expressed in terms of the standard deviation for the forecast variable considered. The standard deviation highlights those regions where the representation of the forecast variable differ most strongly in the individual ensemble members.

The temporal evolution of uncertainty with forecast time is illustrated with an adapted Hovmöller diagram (Hovmöller, 1949) for the standard deviation or spread of a specific forecast variable like in our case the 500 hPa geopotential height. Such a diagram provides insight into the longitudinal varying increase of standard deviation in the ensemble with forecast time. To achieve this temporal-longitudinal pattern, the standard deviation is averaged over a latitude belt between 40°N to 60°N, which typically contains large fractions of the mid-latitude wave pattern that is strongly influenced by the transitioning storm. Those regions where uncertainty grows during the forecast are marked by plumes of standard deviation, whose magnitude and extension increases with time. During an ET for which predictability in downstream regions is reduced, such a Hovmöller diagram typically exhibits a region of enhanced standard deviation in the part of the wave pattern that is directly influenced by the storm. Depending on the strength of the influence it will also propagate in further downstream regions into the subsequent wave pattern as well. By highlighting the position of the transforming tropical storm[1] in the individual ensemble members, the potential impact of the ET event on forecast uncertainty in downstream regions can be investigated as a plume of increased standard deviation, emanating from the positions of the storm.

Such Hovmöller diagrams provide a first indication of the differences in terms of forecast uncertainty between the individual EPS that contribute to TIGGE. Three representative individual EPS and TIGGE already indicate the existence of large differences (Figure 4.1). For the 9-day forecast, initialised 96 hours prior to the ET of Hurricane Ike, the variability in the forecast position of Ike differs among the individual EPS and is larger overall in TIGGE, with some members moving westwards without recurvature and others moving northeastwards and undergoing ET. The strongest uncertainty in the 500 hPa geopotential height field occurs in the region downstream of the transitioning hurricane in only two of the three individual EPS and TIGGE, but the structure and strength of the spread differs. The Australian EPS indicate relatively small variability in the storm positions and less standard deviation downstream therefrom. The variability in storm positions is larger in the Canadian EPS, leading to a western and eastern branch (80°W and 50°W on 17 September) of increased standard deviation. In the ECMWF EPS, this separation can be identified as well,

---

[1] exactly: the point where it has minimum sea level pressure

while the eastern region (50°W on 17 September) causes most of the spread among members. TIGGE clearly provides the largest variability in storm positions, but the pattern of spread is as represented by the individual sets defined above. The reduced absolute values in spread in the TIGGE data set might be explained by the enhanced number of ensemble members. If many members capture the same region of phase space (have a related pdf, Section 2.2.2) the maximum deviation from the ensemble mean is not increased, while the total number of members used to calculate the the standard deviation is.

Related behaviour of forecast uncertainty could be observed in the other cases as well. However, the forecast which was initialised prior to the ET of Hurricane Hanna (Table 4.2) stands out, as the Hovmöller diagram of for the TIGGE data closely resembles the one for the ECMWF data (Figure 4.2). This is a first indication that this forecast has some special properties, as we will see later. This comparison indicates that there are clear differences in variability and storm positions in the individual EPS contributing to TIGGE. These differences suggest a variety of possible development scenarios to be found in the TIGGE data set, maybe more than in an individual ensemble system.

4.3 Variability Captured by EOF Fields

The EOFs allow us to identify those components of the synoptic flow (e.g. flanks of a trough or amplitude of a ridge) which exhibit primary differences among the EPS members at one specific forecast time. This time is chosen to be the first synoptic time (00 UTC, 12 UTC) after which the strongest increase in standard deviation in the downstream region of the transforming storm could be identified in the TIGGE data set. This will expose the reason, i.e. the potential influence of the transitioning storm, for the different development of the forecast scenarios. Investigation at the

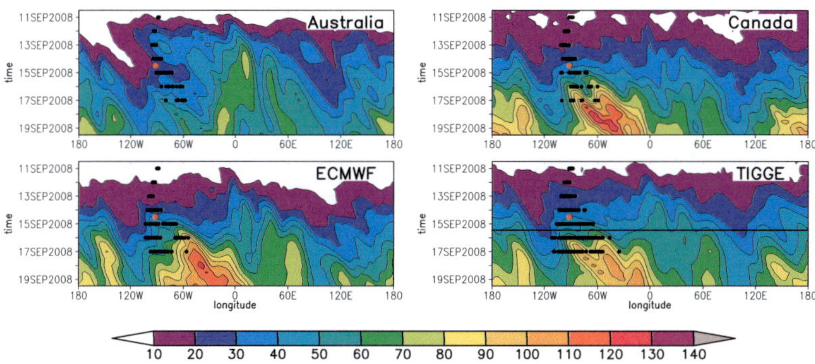

Figure 4.1: Hovmöller diagram for the standard deviation of the 500 hPa geopotential height (in gpm) in the Australian, the Canadian, the ECMWF and the TIGGE EPS for the sample case (Ike10). Surface position of Ike in all ensemble members is marked by the black dots, best track position at ET time by the red dot. Black line in TIGGE panel marks the investigation time (taken from Keller et al., 2011).

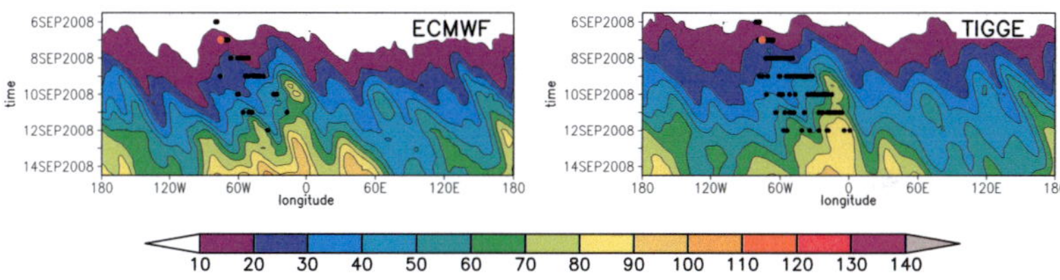

Figure 4.2: As Figure 4.1 but for forecast of Hurricane Hanna, where ECMWF and TIGGE show closely related patterns of standard deviation.

time of maximum variability, which typically occurs several days after ET would only point to the resulting scenarios, thus rather to the symptom than the cause (Harr et al., 2008, Anwender et al., 2008). To make the outcome of the three data sets comparable, the time identified for the TIGGE data set is also chosen as investigation time for the TI-EC and ECMWF data set. For the sample case, this time is marked by the black line in Figure 4.1.

For Ike10, the strongest variability in TIGGE (Figure 4.3 a,d), described by EOF 1 and 2, can be found in the position and amplitude of the trough interacting with Ike, the extension of the downstream ridge towards the north, and in the wave pattern developing further downstream. In TI-EC, the predominant variability occurs mainly in the downstream region: the strongest variability is associated with the amplitude of the downstream ridge and the ridge flank close to Ike, and the second strongest with the ridge position and tilt (Figure 4.3 b,e). This resembles the shift pattern identified by Anwender et al. (2008). The ECMWF members vary most strongly in the flanks of the trough close to Ike followed by the amplitude of the ridge southeast of Greenland (Figure 4.3 c,f). In the sample case, the first two EOFs capture between 33 and 39 % of the total variability contained in this forecast (Table 4.2, Figure 4.3).

How the individual ensemble members contribute to these differences becomes obvious by comparing the geopotential height field of some members to the ensemble mean (Figure 4.4 for Ike10). This shows how the members differ from the ensemble mean and thus contribute to the EOF fields[2]. EOF 1 of the TIGGE data set (Figure 4.3a) is dominated by negative amplitudes in the mid-latitude region and positive amplitudes towards the north. Members with a positive contribution thus exhibit enhanced geopotential heights in northern regions ($60\text{-}70°$N, $40\text{-}80°$W, Figure 4.4a) compared to the ensemble mean while members with negative contribution have lower geopotential heights in these regions (Figure 4.4c). The trough with which Ike is interacting during its ET is broader and has a more pronounced shortwave structure close to Ike in those members with positive contributions to EOF 1 (thus, lower geopotential height in the regions with negative EOF 1 amplitudes, Figure 4.4a). In contrast, this trough is much narrower and has different shape and tilt in negatively contributing members (Figure 4.4c). Finally, the negative EOF 1 amplitude southwest of Iceland (Fig-

---

[2]since the EOFs indicate where the members show strongest variability and thus strongest deviations from the ensemble mean

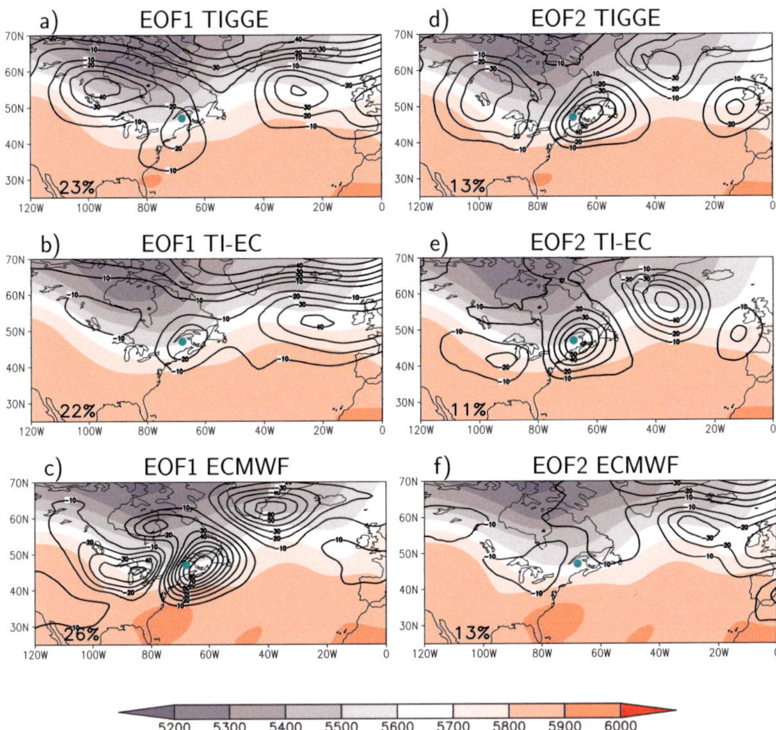

Figure 4.3: Distribution of EOF 1 (left) and EOF 2 (right)(contours, in gpm) and the ensemble mean of the geopotential height at 500 hPa (shading, in gpm) for the TIGGE (top), TI-EC (middle) and ECMWF (bottom) data set for Hurricane Ike forecast, initialised 12 UTC 10 September 2008. Investigation time given in Table 4.1. Percent of variability captured by the EOF is given in lower left corner. At this time, Hurricane Ike (green dot) is located at 45°N, 80°W in ECMWF control forecast (taken from Keller et al., 2011).

ure 4.3a) is caused by positively contributing members that exhibit lower geopotential heights and thus a trough (Figure 4.4a) and members with negative contributions which are characterised by a slight ridge. By comparing the influence of individual members on the second strongest variability in TIGGE, expressed as EOF 2, it is clear that the EOF signals on the flanks of the broad trough interacting with Ike cause a shift of this trough structure, with an eastern trough position in negatively and a western position in positively contributing members (Figure 4.4b,d). Differences in the amplitude of the wave pattern in the downstream region are obvious also. Members with a positive contribution to this EOF distribution have lower geopotential height (negative EOF 2 amplitude, Figure 4.3d) southeast of Greenland and increased geopotential height west of Europe, which leads to a rather zonal flow (Figure 4.4b). In contrast, members with a negative contribution to EOF 2 exhibit a strongly amplified wave pattern in the downstream region, strong ridge (high geopotential heights) south of Greenland and pronounced trough (low geopotential heights) west of Europe (Figure 4.4d).

For all ten EPS forecasts the first two EOFs capture between 26 and 42 % of the total variability with a higher percentage in ECMWF than in TIGGE or TI-EC (Table 4.2). These amounts of total variability are comparable to the findings of Harr et al. (2008) and Anwender et al. (2008). To address the question as to whether the patterns of variability are related to each other in the three data sets in a quantitative manner the similarity index SI is used (Chapter 3). In this framework, SI is applied to the EOFs of two fields, which already state a deviation from the ensemble mean. This supersedes the formation of perturbation fields in the first step and SI can be computed as follows

$$SI = \frac{\mathbf{eof}_m^A \cdot \mathbf{eof}_n^B}{\left|\mathbf{eof}_m^A\right|\left|\mathbf{eof}_n^B\right|} \qquad (4.1)$$

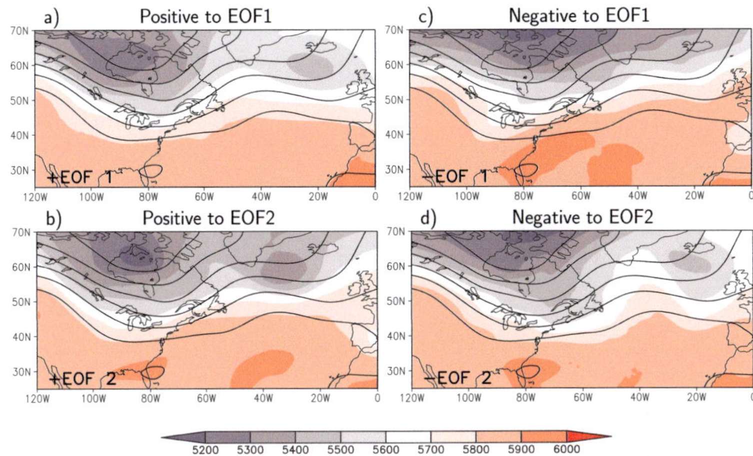

Figure 4.4: Geopotential height at 500 hPa (shaded, in gpm) for several TIGGE members and for the ensemble mean (contour). The members contribute a) positive to EOF 1, b) positive to EOF 2, c) negative to EOF 1, d) negative to EOF 2. Contribution to the other EOF, respectively, is rather neutral.

with **eof** being the state vector of EOF 1 or EOF 2 ($m, n = 1, 2$) of two of the three data sets ($A, B =$TIGGE, TI-EC, ECMWF). The SI was calculated for all four combinations of EOF 1 and EOF 2 for a given pair of data sets. The two highest SI values for the four combinations (SI_1 and SI_2) are given in Table 4.2, with an indication of whether the same EOFs of the data set are being compared (N: $m = n$) or if they are cross related (C: $m \neq n$). The outcome of this quantitative measure for resemblance is then checked by comparison of difference plots between the two EOF fields in question.

A comparison between the EOF distributions for Ike10 (Figure 4.3) indicates that in this case the strongest uncertainty (EOF 1) in the TIGGE and TI-EC data set occurs mainly in the same regions with related amplitudes ($SI = 0.92$, Table 4.2). The EOF 2 patterns differ more strongly in amplitude and regions of uncertainty ($SI = 0.56$).

Table 4.2: Overview of variability in the five ET cases studied (H: Hurricane, TY: Typhoon, TS: Tropical Storm). Left: forecasts and investigation times. Middle: The two largest values of the similarity index SI for the comparison of EOF 1 and EOF 2 between the different data sets, N=normal relation, C=cross relation (see text for details). Right: total variability captured by EOF 1 and EOF 2 (taken from Keller et al., 2011).

Forecast		Similarities of EOFs						% of variability		
ID	Name	TIGGE&TI-EC		TIGGE&ECMWF		TI-EC&ECMWF		TIGGE	TI-EC	ECMWF
1	Hanna05	1.00 N	0.98 N	0.92 N	0.93 N	0.91 N	0.92 N	36	33	38
2	Jangmi26	0.99 N	0.99 N	0.69 C	0.30 C	0.60 C	0.32 C	35	39	34
3	Ike12	0.98 N	0.99 N	0.55 C	0.48 C	0.57 C	0.47 C	41	42	43
4	Bavi17	0.98 N	0.95 N	0.89 N	0.68 N	0.80 N	0.50 N	31	32	40
5	Sinlaku16	0.99 N	0.77 N	0.72 C	0.34 C	0.48 C	0.27 C	26	31	32
6	Bavi18	0.94 N	0.77 N	0.64 N	0.33 N	0.64 N	0.28 N	30	30	40
7	Ike11	0.81 N	0.85 N	0.89 N	0.37 N	0.81 N	0.41 N	35	30	41
8	Sinlaku15	0.96 N	0.42 N	0.44 N	0.33 N	0.21 C	0.21 C	30	31	30
9	Sinlaku14	0.96 N	0.34 N	0.60 C	0.16 C	0.57 C	0.50 C	29	30	32
10	Ike10	0.92 N	0.56 N	0.78 C	0.77 C	0.81 C	0.69 C	36	33	39

These findings are confirmed by the difference plots for the EOF 1 and EOF 2 distributions in the TIGGE and TI-EC data sets (Figure 4.5). The TIGGE/TI-EC are cross-linked to those of the ECMWF data set. Thus, $SI_{1,2}$ between TIGGE EOF 1 (EOF 2) and ECMWF EOF 2 (EOF 1) is 0.78 (0.77) and 0.81 (0.69) between TI-EC and ECMWF, respectively (Table 4.2). A generalisation to all ten cases (Table 4.2) allows to identify typical relationships between the pattern of variability. In case 1, the forecast that was initialised prior to the ET of Hurricane Hanna, the EOFs are nearly similar in all three data sets. This indicates that the data sets show closely related patterns of variability, or, in other words, the individual members of the three data sets differ similarly from their respective mean state. In cases 2, 3 and 4, the EOF distributions are closely related in the TIGGE and TI-EC data set ($SI_{1,2} > 0.95$ and absolute differences less than 10 gpm (not shown)), but not in the comparison with ECMWF (although in case 4 $SI_1 > 0.8$ in the comparison with ECMWF). This high similarity indicates that the variability in the ECMWF does not dominate the strongest and second strongest (EOF 1 and EOF 2) variability in the TIGGE data set, as an exclusion of the ECMWF data (resulting in the TI-EC data set) does not remarkably change the patterns of uncertainty. In the remaining forecasts, SI_1 is high (>0.78) between all ensemble sets (cases 4, 7, 10) or, at least does not drop below a value of 0.8 (cases 5, 6, 8, 9) between the TIGGE and TI-EC set. On the other hand, SI_2 is significantly lower in almost all of these cases. Thus, ECMWF seems to influence the second strongest variability in the TIGGE data set, as its exclusion causes differences in the EOF 2 distribution of the TI-EC data set. In approximately half of the cases, the closest resemblance is found for cross-linked EOFs (EOF 1 of TIGGE/TI-EC and EOF 2 of ECMWF and vice versa). This indicates that the differences among the members in the TIGGE/TI-EC and ECMWF data sets occur at least in the same regions, but that they have a various contribution to the overall variability in the forecast.

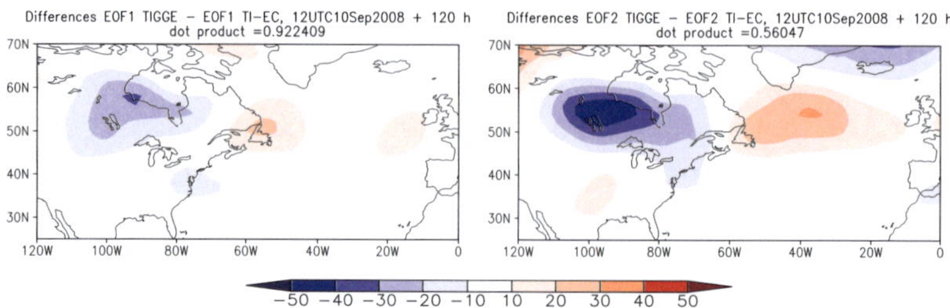

Figure 4.5: Differences between EOF 1 (left) and EOF 2 (right) in the TIGGE and
TI-EC data set for the sample case (taken from Keller et al., 2011).

Overall, this comparison shows strongest and in particular second strongest differences between the individual TIGGE members occur in other regions than between the ECWMF members (rather low similarity indices between TIGGE/TI-EC and ECMWF in many cases), but that ECMWF exhibits patterns of variability not seen in TI-EC, as the exclusion of ECMWF causes differences.

4.4 Contributions of Individual EPS to TIGGE Clusters

These distributions of uncertainty arise due to a superposition of the different variability patterns in the eight TIGGE EPS. The manner in which the members of the individual EPS contribute to this variability can be investigated by using the clustering results. Although the clusters are distributed over the whole phase space (PC1, PC2) spanned by EOF 1 and EOF 2, some of the TIGGE EPS only capture parts of the overall variability and thus contribute to just some of the clusters. This can be expressed by the percentage number of members from the several EPS (y-axis) assigned to a given cluster (colours, Figure 4.6). For Ike10 Brazil contributes to only one cluster (cluster 6), but this cluster has contributions from five other EPS also, thus it is not an outlier formed by only one EPS. Australia contributes to only two of the six identified clusters (clusters 4, 5), while members from the Japanese and ECMWF EPS contribute to five of the six clusters. Thereby, Australia mainly shares clusters (i.e. shares colours) with the EPS from Canada, UKMet and Japan. All clusters have contributions from four or more EPS, except for cluster 3, that is dominated by the ECMWF. The clustering result of all ten forecasts is investigated in the same manner (Figure 4.7). The lowest level of contribution to different clusters is seen for the Brazilian EPS, contributing mainly to one cluster. The Australian, Chinese, NCEP and UKMet EPS are characterised by having most of their members assigned to one or two clusters and only a few in other clusters. A stronger separation into different clusters occurs for the Canadian and ECMWF EPS. The Japanese EPS is notable for contributing to almost all clusters in each forecast. Furthermore, the ECMWF EPS often dominates one cluster (cases 6, 7, 9, 10). Thus, ECMWF seems to introduce a new cluster, i.e. a new possible development scenario into TIGGE. Figure 4.7 indicates also which EPS contribute to the same clusters (sharing colours). It appears that particular groupings occur in a significant proportion of the forecasts. To quantify this we calculate the number of cases in which two of the EPS considered share at least 20% of their members (Table 4.3). The Brazilian and to a lesser extent the Australian EPS seldom share clusters with other EPS systems. The Canadian EPS shares clusters with nearly all of the other EPS, and in particular, shares at least one cluster with the members from Japan in all ten cases. An overall weak mixing of members between clusters was found in case 5 and might be due to the initialisation of the forecast being closer to the ET time of Sinlaku. The same analysis was repeated for the TI-EC data set to investigate whether the cluster contributions are changed by taking out the ECMWF data set. Although some details changed (Table 4.4), now NCEP and Canada share clusters in all ten cases whereas the number of cases with shared members dropped slightly, the overall behaviour is confirmed.

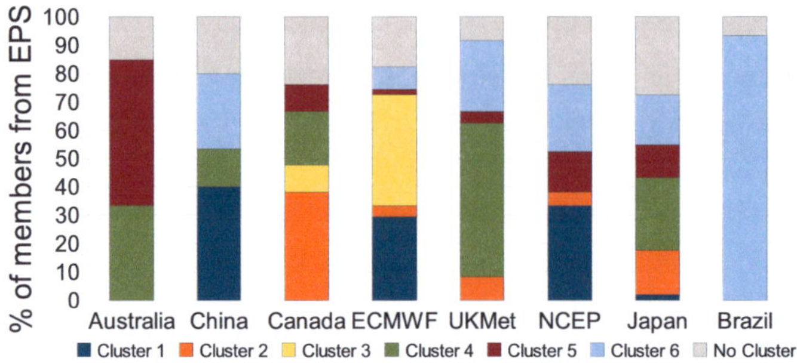

Figure 4.6: Contribution of members from the individual EPS to the different clusters (in colours) for Ike10. Ordinate: % of members from the given EPS contributing to the cluster in question (in colour). Members that could not be clearly assigned to one cluster are outlined in grey (taken from Keller et al., 2011).

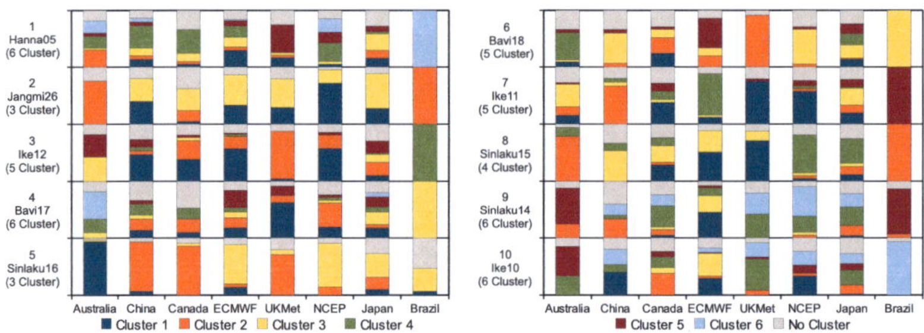

Figure 4.7: As Figure 4.6 but for all ten cases. The bottom right panel (Ike10) is reproduced from 4.6 for completeness (taken from Keller et al., 2011).

Table 4.3: Number of cases in which at least 20% of the members from the EPS share the same cluster in the TIGGE data set, orange:< 5 cases, yellow: ≥ 5 and < 8 cases, green:≥ 8 cases (taken from Keller et al., 2011)

	Australia	China	Canada	ECMWF	UKMet	NCEP	Japan	Brazil
Australia		3	3	1	1	2	5	4
China	4		8	6	6	7	7	2
Canada	3	8		8	9	8	10	0
ECMWF	1	6	8		6	7	6	1
UKMet	1	6	9	6		8	8	1
NCEP	2	7	8	7	8		9	4
Japan	5	7	10	6	8	8		3
Brazil	4	2	0	1	1	4	3	

Table 4.4: Same as Table 4.3, but for the TI-EC data set.

	Australia	China	Canada	UKMet	NCEP	Japan	Brazil
Australia		4	4	4	4	6	6
China	4		9	6	9	8	3
Canada	2	9		9	10	7	1
UKMet	2	6	9		7	6	0
NCEP	3	9	10	7		8	3
Japan	6	8	7	6	8		3
Brazil	6	3	1	0	3	3	

The conclusion from this analysis is that typical patterns of behaviour can be identified for the contribution of the individual EPS to ET scenarios and for the groupings within the scenarios. Neither the degree of variability in the individual EPS (contributions to cluster) nor the characteristic groupings appear to be related to the initialisation method. For example, the JMA and Canadian or NCEP EPS share one major cluster in nine or even all of the ten forecasts, but use different methods to create the initial perturbations for the ensemble.

4.5 Differences in the Scenarios

To pursue the question as to whether different development scenarios are obtained for TIGGE, TI-EC and ECMWF, we investigate the different development scenarios in spaghetti plots for the cluster mean of the 500 hPa geopotential height. For Ike10 (Figure 4.8), two ECMWF clusters (blue, both panels) show a shift in the wave pattern with a trough at 50°W (right box), a slight ridge northeast of the Great Lakes (left box) and ridging west of the British Isles +48 hrs after the investigation time. Members of these clusters contribute to a similar and a related TIGGE scenario (red, top panel) that are not present in TI-EC (black, lower panel). One of the ECMWF clusters has a strongly amplified ridge west of Europe, which is found in TIGGE only to a lesser extent, as the grouping of these ECMWF members with other TIGGE members smoothes the amplified ridge in the cluster mean. The regions where the scenarios diverge correspond nicely with the regions of enhanced uncertainty in the TIGGE Hovmöller diagram (Figure 4.1) on 17 September2008, 12 UTC, i.e. 90°-60°W (left box), 55°-30°W (right box) and 30°-0°W (ridge flank west of Europe). Here, the use of TIGGE instead of ECMWF would provide four different scenarios compared to the use of ECMWF alone, but using TI-EC would result in two missing scenarios. Thus, TIGGE as a whole seems to offer the largest range of possible solutions, but only if the ECMWF is included.

The agreement between the TIGGE and TI-EC data sets is further quantified using the Rand Index RI (Section 3.1.5), which simply measures the numbers of members that are assigned likewise during the clustering of the two data sets under investigation. A RI close to 1 indicates, that the two cluster solutions are nearly identical, while RI is zero if all members are assigned differently in the two cluster solutions. In order to compare TIGGE and TI-EC, the TIGGE clustering was carried out with the ECMWF members, but these were removed from the clusters before calculating the RI. In addition to this quantitative measure of similarity, the synoptic structure of the cluster means of 500 hPa geopotential height is compared using spaghetti plots as well as synoptic maps of the 500 hPa geopotential height and mean sea level pressure. The clusters are then categorised in three groups, for high, moderate and low agreement (Table 4.5), depending on the differences in their geopotential height pattern around investigation time. High agreement is achieved if all TIGGE and TI-EC show nearly the same scenario in the spaghetti plots; moderate, if they show some differences but overall behaviour is confirmed; low if the respective scenarios differ clearly in parts of the flow (Figure 4.9). In this framework, forecasts 2 and 3, which also showed related EOF distributions (Table 4.2) turned out to have nearly the same development scenarios in TIGGE and TI-EC ($RI > 0.89$, Table 4.5). This confirms that the variability in the TIGGE data set does not change when the ECMWF EPS is disregarded. However, case 1 is conspicuous in this investigation, as its standard deviation and EOF distributions indicated close resemblance between all three data sets. Here it turns out that the cluster result and thus the developments scenarios differ more than in other cases. This indicates that the members might have a slightly different distribution in the phase space, causing the development of distinct cluster solutions, i.e. unassigned members in one data set will be assigned in the solution for the other data set.

With a decreasing *RI*, the number of closely related clusters in the TIGGE and TI-EC data set decreases also and the cluster solutions start to differ more strongly. This is consistent with the decreasing similarity of the EOF patterns (Table 4.2). The TIGGE or TI-EC and ECMWF clusters, respectively, exhibit only moderate agreement overall, but, as expected, the agreement is stronger for those cases where the TIGGE and TI-EC forecasts were related. This might be due to the fact that the ECMWF clusters often contain a smaller amount of members and smaller scale features are not as smoothed out as they are in the TIGGE clusters with large memberships.

4.6 Poor Man's Verification

An aim of this study is to determine whether we get more possible development scenarios using TIGGE instead of ECMWF. Thus it is important to comment on some aspects of forecast uncertainty and its verification. We have seen that more possible scenarios often exist in TIGGE than in ECMWF. This does not necessarily mean that the TIGGE data set will give the best possible ensemble forecast. The variability captured in an ensemble forecast should not exceed the real uncertainty of the future atmospheric state, i.e. the ensemble should not show a large variability, when it is nearly certain how the atmospheric development will continue over the time frame of the forecast. Thus, in some cases, the variability found in the TIGGE

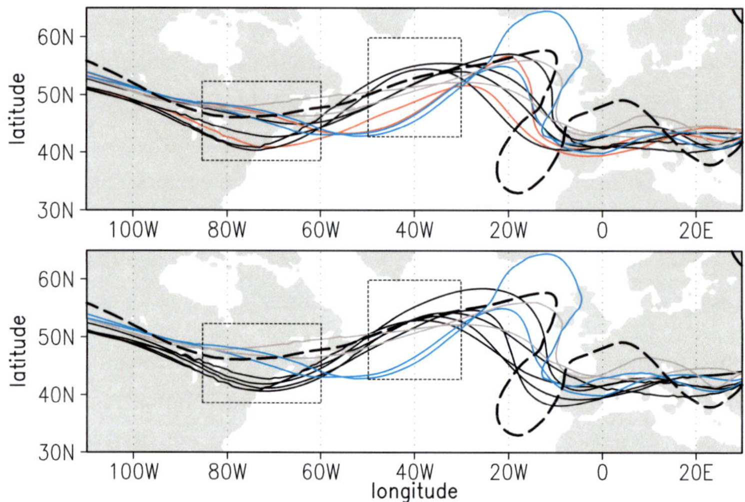

Figure 4.8: Different development scenarios for Ike10, 48 h after investigation time (12 UTC 17 September 08, verifying analysis dashed). Spaghetti plots of the 570 gpm isoline cluster means. Top: TIGGE (3 black, 2 red) and ECMWF (blue), bottom: TI-EC (black) and ECMWF (as above). Colours to highlight scenarios mentioned in text. Boxes discussed in text (taken from Keller et al., 2011).

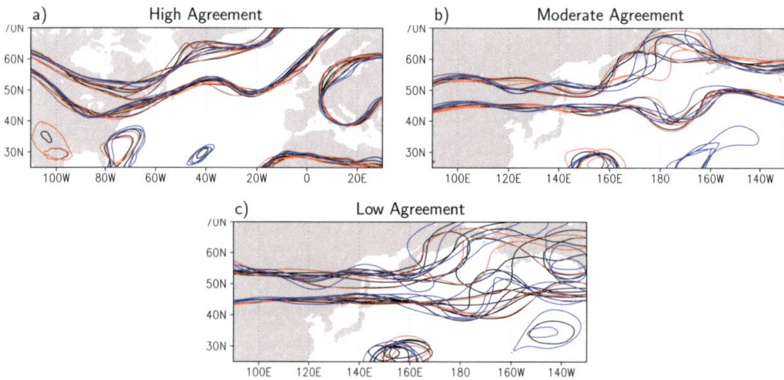

Figure 4.9: Spaghetti plots for 5500, 5700, 5900 gpm isolines of cluster means from the TIGGE (black), TI-EC (red) and ECMWF (blue) development scenarios. Shown are three cases, which characterise a) high, b) moderate and c) low agreement between the black (TIGGE) and red (TI-EC) lines.

Table 4.5: Similarity of clusters in different EPS sets. Rand Index between the clusters for the TIGGE and TI-EC data sets (column 2) and number of clusters in each forecast showing high, medium and low agreement for TIGGE compared with TI-EC (column 3), TIGGE with ECMWF (column 4) and TI-EC with ECMWF (column 5), in terms of resemblance of development scenarios (taken from Keller et al., 2011).

			TIGGE & TI-EC			TIGGE & ECMWF			TI-EC & ECMWF		
ID	Case	Rand Index	high	mod.	low	high	mod.	low	high	mod.	low
2	Jangmi26	0,92	3	–	–	–	2	1	–	2	1
3	Ike12	0,90	4	1	–	–	3	2	–	2	3
1	Hanna05	0,86	2	3	–	–	3	2	–	2	3
4	Bavi17	0,82	–	2	4	–	2	4	–	1	5
6	Bavi18	0,81	3	2	–	–	2	3	–	1	4
5	Sinlaku16	0,81	1	1	2	–	1	2	–	1	3
7	Ike11	0,78	–	2	3	–	1	4	–	–	5
8	Sinlaku15	0,77	–	1	3	–	–	4	–	–	4
10	Ike10	0,77	–	1	5	–	1	5	–	–	6
9	Sinlaku14	0,72	–	1	4	–	–	5	–	–	5

Table 4.6: Percentage number of members that contribute to the cluster which resembles the analysis most closely.

	Case	Australia (33)	China (15)	Canada (21)	ECMWF (51)	UKMet (24)	NCEP (21)	JMA (51)	Brazil (15)
1	H Hanna05	21	40	43	20	0	33	4	0
2	TY Jangmi26	0	40	5	33	29	71	27	0
3	H Ike12	0	47	38	57	4	57	10	0
4	TS Bavi17	0	7	0	31	17	5	18	0
5	TY Sinlaku16	97	7	5	82	8	76	51	47
6	TS Bavi18	6	7	5	53	0	5	18	0
7	H Ike11	0	7	14	73	0	5	2	0
8	TY Sinlaku15	94	13	19	0	0	71	57	100
9	TY Sinlaku14	64	0	0	4	0	0	0	80
10	H Ike10	0	40	0	29	0	33	2	0

data set might be too large, compared to the real uncertainty of the future state of the atmosphere. In the framework presented here, it is not possible to construct a verification in terms of how the variability in the forecast captures the real uncertainty of the atmospheric state. This would only be possible if a large number of cases and forecasts were investigated that allow a statistical investigation of the uncertainty, i.e. it would be necessary to investigate how often a specific feature occurred in the analysis and how often it did not in all the investigated cases. Then, the comparison of this frequency of occurrence could be compared to the variability and frequency of occurrence of the event in the forecasts. If these frequencies of occurrence would be related to each other, the forecast would be reliable and give a correct representation of the uncertainty in the atmospheric development.

A subjective verification can be performed by translating a verifying analysis for the investigation time into the PC1-PC2 phase space (by projection, von Storch and Zwiers, 1999). In this way, the verifying analysis can be displayed in the phase space that is used to determine the clusters. The analysis is not considered for the clustering procedure to prevent modifications in the cluster assignments. Examination of the scatter plot for the clustering solution then reveals which cluster resembles the analysis most closely (smallest distance between analysis and cluster centre). This is done for the associated analyses[3] of all eight contributing centres. Thus, eight additional points occur in the scatter plot. In nine of the ten cases, the analyses from the different centres are strongly related and thus lie close together or even coincide in the scatter plot. In case 2, the forecast for Typhoon Jangmi, the verifying analysis from China differs from the other ones and is located in another part of the phase space, but is still assigned to the same cluster as the analysis. To quantify how well the individual EPS show scenarios that are related to the analysis, the percentage number of members of the constituent EPS that are assigned to the cluster which

---

[3]Thereby, the first time step of the control forecast, initialised at investigation time is used as verifying analysis

best resembles the analysis is determined (Table 4.6). If the analyses are located in boundary regions between two clusters, both of these clusters are regarded.

From this comparison (Table 4.6) it is obvious that a number of the EPS (ECMWF, China, JMA, NCEP and Canada) share clusters with the analysis with at least one of their members in nearly all cases, while others (Australia, UKMet and Brazil) often have none of their members in the cluster which best resembles the analyses. However, in two cases (8, 9) the Australian and Brazilian models do a very good job and have a large amount of their members assigned to the analysis related cluster, while most of the other EPS fail to predict the correct development scenario. This indicates that those EPS, although they often do not contain the analysis, can produce realistic development scenarios.

4.7 Discussion and Conclusions

In this part of the study, the variability in the forecast synoptic patterns associated with ET events in TIGGE is investigated using a combined EOF- and clustering analysis. Three main questions were addressed, which enable to highlight the characteristic behaviour of the data set during the analysis methodology. One question was whether the combination of the several EPS in TIGGE provides increased representation of forecast variability and, if it does, whether increased variability allows a broader range of possible development scenarios to be identified in TIGGE rather than in a single model EPS. A further question was whether the individual constituent EPS contribute to this variability and if some of them span the whole phase space of possible development scenarios, while others mainly focus on specific parts of it. To answer these questions, this study is based on the investigation and comparison of three different data sets: TIGGE as a whole, TIGGE without ECMWF (TI-EC) and ECMWF alone.

The outcome of this study shows that the individual EPS exhibit different representations of forecast variability associated with the ET event and storm positions (Hovmöller diagrams, Figure 4.1) in nearly all of the cases. Thereby, the forecast for Hurricane Hanna (case 1) stands out clearly, as its distribution of standard deviation is closely related in the TIGGE and ECMWF data set. The differences in forecast variability in the individual EPS contributing to TIGGE also cause the regions where the members differ most and second most strongly (EOF 1 and EOF 2) to differ between the several data sets (Table 4.2). A comparison of the EOF distributions in the individual data sets, using a similarity index, shows that these variability patterns are nearly identical in the particular forecast case of Hanna. Thus, the forecast variability explained by TIGGE and ECMWF (and also TI-EC) is nearly similar and one would expect closely related development scenarios as well. In three cases, the EOF patterns in TIGGE and TI-EC are closely related, but those in the ECMWF data set differ. From this, we can conclude that the ECMWF EPS explains other variability patterns than TIGGE, but that this variability is non-dominant in TIGGE, as the exclusion of ECMWF does not change the EOF pattern of the TIGGE data set (TI-EC) significantly. In the remaining six cases only the strongest variability (EOF 1) shows related patterns in TIGGE and TI-EC, while the EOF 2 pattern differs more strongly between the TIGGE and TI-EC data set. Here, ECMWF seems to influence the second strongest variability of the TIGGE data set, as its exclusion causes differences. Overall, ECMWF often shows distinct variability patterns (low *SI*), whose importance (strongest or second strongest) also differ from the ones in TIGGE/TI-EC in approximately half of the cases.

As these results already indicate, differences also exist in the clustering solutions. According to the differences in standard deviation, the individual EPS contribute differently to the variability patterns and thus have distinct contributions to the clusters. Some EPS (Japan and ECMWF) contribute to a large number of clusters and thus show a variety of possible development scenarios, while other EPS (Brazil, Australia) only contribute to few clusters and mainly exhibit one development scenario. ECMWF often turns out to dominate one cluster and thus seems to introduce a new development scenario into TIGGE. Furthermore, we could identify preferred groupings of the individual EPS. These results seem to be independent of the initialisation method of the EPS. A comparison of the resulting cluster means, i.e. development scenarios indicate that the scenarios found in TIGGE and TI-EC are closely related in those cases, where their EOF distributions resemble each other. Surprisingly, in the special case of Hanna, the cluster assignments and thus the development scenarios are slightly modified between the TIGGE and TI-EC clustering solution. This is caused mainly by members that are not assigned to the clusters in one data set, but are assigned in the other. Thus, the clustering results show a slight shift in the PC1-PC2 phase space. In those cases, where the EOF 2 distributions are different, the clustering solutions between TIGGE and TI-EC start to differ also. Finally, as the patterns of variability often highlight different regions in the ECMWF to the TIGGE and TI-EC data sets and as ECMWF dominates one cluster, only some of the ECMWF development scenarios are contained in TIGGE. The others are missing and thus do not occur in the TI-EC data. Hence, ECMWF seems to introduce an additional development scenario into TIGGE.

To conclude, TIGGE exhibits more variability and thus offers a broader range of possible development scenarios during an ET event than ECMWF alone. Thus, if the verifying analysis is not covered by the phase space of the synoptic developments found in the ECMWF EPS, we might find a verifying scenario in TIGGE. However, in more than half of the cases studied, the presence of the ECMWF members in TIGGE is necessary to obtain the full range of possible scenarios mentioned above. The additional scenarios resulting from TIGGE provide a means to investigate the physical and dynamical processes responsible for the variability among EPS members and to connect these mechanisms to the predictability of the weather events investigated. This closer investigation of the underlying processes forms the second part of this thesis. Using an analysis of the eddy kinetic energy, the influence of the transitioning storm can be examined. Ensemble sensitivity analysis (Torn and Hakim, 2009) will provide insight into connections between the development of two or more representative variables. Another possibility, provided by the methodology for analysing TIGGE data presented here, is to use the main development scenarios as initial- and boundary conditions for high resolution regional ensembles and thus quantify predictability on smaller scales (e.g. associated with convection) in more detail. This pre-selection might allow specification of a broad range of variability in regional ensembles even with a limited number of members. Finally, this study points to the utility of the TIGGE data set and to the need for continued analysis of information contained in the combinations of EPS from various operational systems.

4.8 This Study in the Framework of other TIGGE Investigations

After TIGGE was initiated in 2005, the contributing centres successively started to archive their ensemble forecasts in the TIGGE archive. Since February 2008, the full set of ten ensemble forecasts is available. In the present study, the EOF- and cluster analysis method is applied to the TIGGE data set to investigate forecast variability during ET events for the first time. At the time of writing this work stands out in other recent studies based on TIGGE data, as they focused mainly on the investigation of different forecast skill scores for several forecast variables and on the best combination and calibration of the individual ensemble system. Here, the main focus is put on the investigation and comparison of synoptic fields and variability in forecast development scenarios.

Park et al. (2008) investigated 500hPa geopotential height and 850 hPa temperature forecasts from some of the constituent TIGGE EPS and verified them against their own analyses. Furthermore, they combined the individual EPS to form a multimodel data set and applied a bias correction. They found the multimodel EPS to have a weak benefit against the bias corrected ECMWF single model EPS for the 500 hPa geopotential height forecasts on the northern hemisphere in terms of the Root Mean Square Error (RMSE) of the ensemble mean and the Ranked Probability Skill Score (Wilks, 1995). On the other hand, they found a larger benefit of the multimodel EPS in forecasts for the tropical 850 hPa geopotential height. These findings coincide with a study from Matsueda and Tanaka (2008) who compared the benefit of a non-weighted and non-calibrated multimodel EPS forecast for the 500 hPa geopotential height in various seasons against the ECMWF single-model EPS using the same verification measures as Park et al. (2008). Therein, the multimodel EPS outperforms the ECMWF EPS in the medium forecast range in all seasons for several percentage points. Matsueda and Tanaka (2008) recommend weighting the individual EPS differently and applying a bias correction during calibration to improve the performance of the multimodel EPS. Especially surface components, like 2m temperature or wind field might even benefit more strongly from calibration than the mid- and upper tropospheric wind field. However, investigation of adequate calibration parameters by Johnson and Swinbank (2009) showed even the raw multimodel approach to have a better skill than every bias corrected single model EPS, especially for the 2m temperature, a surface variable. Thus, they assigned the calibration of ensemble forecasts a weaker impact on the predictive skill than the simple combination of several EPS to form a multimodel EPS.

Based on these early results, Bougeault et al. (2010) summarised the benefit from the multimodel approach to be large for forecasts of surface- or "weather"-related variables (e.g. 2m temperature, precipitation), but to be weaker for the larger-scale variables, like 500 hPa geopotential height or mean sea level pressure field, as those variables are already consistently forecasted in the contributing EPS and are independent from physical parametrisations. Furthermore, Bougeault et al. (2010) highlighted the improvement in predictive skill to be large if EPS with related skill are combined, while the combination of EPS with high and low forecast skill will cause a weaker improvement. Majumdar and Finocchio (2010) investigated how ensemble forecasts from multimodel EPS can be used to provide probabilistic forecasts for tropical cyclone tracks. Although the individual ECMWF EPS already showed accurate mean and probability forecasts, they expected the quality of tropical cyclone track forecasts to be improved further by the use of a multimodel approach. They argued that the different EPS develop distinct ranges of possible developments (i.e. span different pdfs). This might be confirmed by the findings of the study in hand. The prediction of extratropical cyclones (cyclone position, intensity and propagation speed) in the northern hemisphere in the individual TIGGE EPS was investigated by Froude (2010). They found the ECMWF EPS to have highest level of forecast skill for all cyclone parameters under investigation. The other EPS tend to show enhanced error growth and to be underdispersive.

These results indicate that the multimodel approach leads to an improvement in forecast skill, even for the 500 hPa geopotential height field, which is typically well predicted in many forecast models (Park et al., 2008). This is another perception on the TIGGE data set than it is provided by the study in hand. However, the findings of increased variability and more possible development scenarios in the TIGGE rather than the ECMWF EPS seem to fit in the frame spanned by other TIGGE-studies.

5 Investigation of Extratropical Transition Using the Eddy Kinetic Energy Budget

The second main part of the present study addresses the more detailed investigation of mechanisms determining the impact of a transitioning tropical cyclone on the mid-latitude flow. Analysis of the eddy kinetic energy budget was already shown (Chapter 3) to provide useful insight into the examination and downstream propagation and development of wave patterns in the mid-latitudes, i.e. Rossby wave trains. Harr and Dea (2009) furthermore gave an example on how the eddy kinetic energy budget can be employed to study the processes of possible reintensification and downstream development during the ET of tropical cyclones.

In this chapter, the eddy kinetic energy budget is analysed for different synoptic developments, which resulted from the application of the EOF- and cluster analysis to ensemble forecasts. Furthermore, the possibility of using the eddy kinetic energy itself as clustering variable is explored. The EKE analysis requires the vertical velocity (ω) as input variable, which is not available in the TIGGE multimodel data set. As vertical velocity is available in the operational ECMWF EPS, at least at some pressure levels, data from this ensemble is used for this part of the study. The availability of the data at only some pressure levels means that the output of the ensemble forecasts for vertical velocity is stored at only these six pressure levels, while the forecast itself is running at a comparatively high vertical resolution. The number of pressure levels, for which the ensemble forecast data is available, is referred to as vertical output resolution in the remainder. Other recent studies typically examined data sets with higher vertical output resolution than is provided by the operational ECMWF EPS. Hence, they could resort to more information about the vertical flow structure. Thus, before applying the EKE analysis to the ECMWF EPS, the question is considered as to whether the coarser vertical output resolution of the ECMWF operational EPS, especially for vertical velocity, is still sufficient to get a consistent picture of processes involved in downstream baroclinic development and during extratropical transition. Thereafter, the EKE analysis will be applied to the appropriate data set to provide the base for a subsequent cluster analysis. A closer examination of the eddy kinetic energy budget for several forecast scenarios then reveals the processes that cause the different developments.

5.1 Influence of Vertical Output Resolution

While the horizontal resolution of the EPS (0.5 ° lat/lon) is comparable to or even higher than in other recent studies, the available pressure levels in the ECMWF EPS are not. Previous studies (e.g. Chang (2000), Danielson et al. (2004), Decker and Martin (2005), Harr and Dea (2009)) used 14 to 20 pressure levels with increments less or equal 100 hPa. Data from the operational ECMWF EPS is only available at nine pressure levels with their increments varying between 50 and 200 hPa. The vertical velocity is available at only six pressure levels, with limited data in the upper troposphere (up to 300 hPa increment). One exception is the conceptual study by Orlanski and Sheldon (1995), who revised their findings using analyses from operational NMC[1] analyses with 10 pressure levels and increments, varying between 50 and 200 hPa. In Glatt et al. (2011), the EKE analysis was applied to the ECMWF operational analysis with 0.5° horizontal resolution and 14 pressure levels (Table 5.1), which matches with other studies.

---

[1]National Meteorological Centre, USA, today incorporated in NCEP

However, as a part of his PhD thesis, Simon Lang (Lang, 2011) used a newly configured setup[2] of the ECMWF EPS to rerun some forecast for tropical storms in the 2008 season. In these new forecasts all variables, including vertical velocity, are available at ten pressure levels. Hence, due to the larger number of available pressure levels, especially for vertical velocity, this ensemble data set (referred to as experimental EPS in the following) could provide a better data base for EKE analysis than the operational ECMWF EPS. However, as this data set is only available for a limited number of forecasts in 2008 it constrains the range of ET events that can be investigated. To possibly overcome this restriction, the influence of the coarse vertical output resolution is explored to examine whether the operational vertical output resolution is sufficient for an investigation of the underlying processes.

5.1.1 Setup for Comparison

The extent to which the different vertical output resolution influence the outcome of the EKE analysis is investigated using the operational ECMWF analysis data set. Based on this data set and its 14 vertical levels, the vertical output resolution of the experimental and operational ECMWF EPS are emulated. The eddy kinetic energy budget for the analysis data set is computed three times, using 1) all available 14 levels, 2) only the ten levels that are available in the experimental EPS data set and 3) only the levels, available in the operational ECMWF EPS. An overview of the pressure levels of different data sets is provided by Table 5.1. The comparison is done for the first 15 days of September in 2008, during which the extratropical transition of Hurricane Hanna could be observed. This is one of the cases to which the EKE analysis is applied.

---

[2]The configuration, described in Table 2.1, extended by an analysis ensemble and a kinetic energy backscatter scheme, Section 2.2.2

Table 5.1: Overview of pressure levels (in hPa) that are available in the distinct data sets. Left: operational analysis, middle: experimental EPS, right: operational EPS and its vertical velocity

operational analysis	experimental ECMWF EPS	operational ECMWF EPS	vert. vel. oper. ECMWF EPS
100	100	100	-
150	-	-	-
200	200	200	200
250	-	250	-
300	300	300	-
400	400	-	-
500	500	500	500
600	600	-	-
700	700	700	700
800	-	-	-
850	850	850	850
900	-	-	-
-	925	925	925
950	-	-	-
1000	1000	1000	1000

For each of the budget computations, the vertical levels listed in Table 5.1 are considered. Vertical velocity affects only the baroclinic conversion term. Hence, for the operational data set this term is computed just for the available six levels, while the other budget terms are computed at all nine levels. Data in the lower troposphere is available at 950 hPa in the operational analysis and at 925 hPa in the EPS data sets. As the budget terms are vertically integrated, data at 950 hPa instead of 925 hPa is used. Vertical velocity is not available at 100 hPa in the operational EPS, where this quantity turns out to be about one magnitude smaller at 100 hPa than at all other levels and hence can be neglected anyway.

5.1.2 Sensitivity of Budget Terms to Vertical Output Resolution

The effect of different vertical resolutions can now be demonstrated by comparing the resulting eddy kinetic energy budget terms for the operational analysis (referred to as "analysis" hereafter) with the two calculations with reduced vertical output resolution (referred to as "experimental" and "operational"). The budget terms of Equation 3.21 and Equation 3.24, computed for the different data sets, are first compared for one specific time. As the EKE analysis will be used to investigate underlying physical processes during ET, the time for comparison is chosen to be the 7 September 2008, 00 UTC, at which Hurricane Hanna started to interact with a preexisting trough in the mid-latitudes during its extratropical transition. Hanna was declared to be an extratropical system 6 hours later.

Overall, the distribution of K_e and of most of the budget terms in the reduced vertical resolution resemble the patterns identified in the analysis data set quite well (Figure 5.1a,b). Differences between the data sets mainly affect the magnitude of the distinct quantities but not their horizontal distribution. The maxima in the three data sets coincide in their horizontal region, i.e. no regions exist, where one data set indicates strong budget terms, while the other data set does not. Interestingly, most of the quantities whose computation does not require vertical velocity (K_e, *generation*, *div-ageo*, *div-keflux*, *brtr* in Table 3.1, p. 70) are closer to the analysis in the operational output resolution than in the experimental. (Figure 5.1a,c,g,i). Thus, although ten pressure levels are used in the experimental output resolution, the operational output resolution with its nine levels seems to resemble the analysis data more closely.

Differences between the experimental and operational resolution are found in the mid- to upper troposphere. While the experimental data set is available in 100 hPa increments above 700 hPa, the operational EPS is output at coarser output resolution in the mid troposphere, but at higher in the

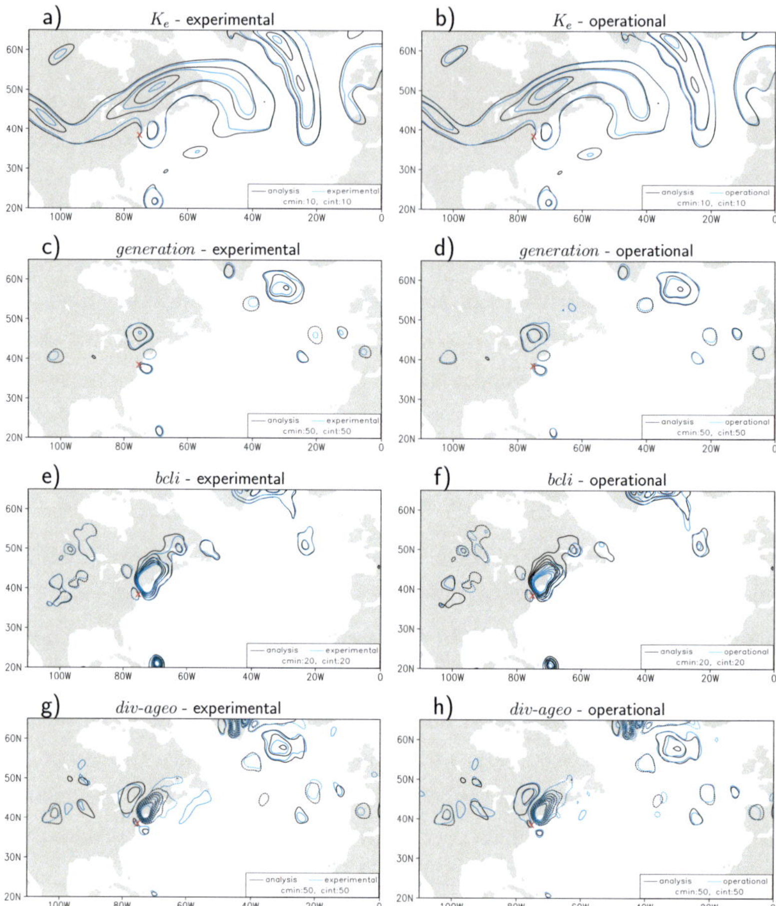

Figure 5.1: Vertically integrated eddy kinetic energy budget terms for 7 September 2008, 00 UTC in analysis output resolution (black contours, both columns), experimental EPS output resolution (blue contours, left column) and operational EPS output resolution (blue contours, right column). Surface pressure centre of Hurricane Hanna is marked by red cross. Contour intervals listed in legend, cint means contour interval, cmin stands for lowest contour plotted. a,b) eddy kinetic energy (K_e), c,d) total generation (*generation*), e,f) baroclinic conversion (*bcli*), g,h) ageostrophic geopotential flux convergence (*div-ageo*), i,j) convergence of K_e flux with total wind (*div-keflux*), k,l) barotropic conversion (*brtr*). K_e in $10^5 \, J/m^2$, budget terms in W/m^2. Continued on next page.

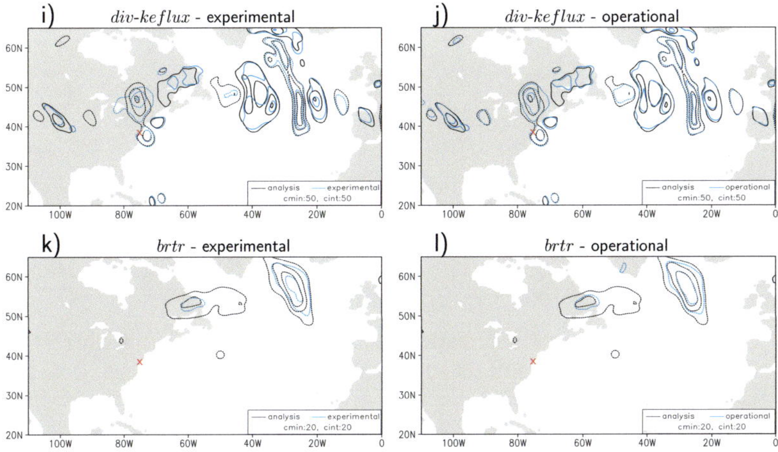

Figure 5.1: continued.

upper troposphere. This enhanced number of pressure levels in the upper troposphere (250 hPa) in the operational output resolution seems to affect the vertically integrated velocity-dependent budget terms and K_e more than the additional mid-tropospheric pressure levels (600 hPa, 400 hPa) in the experimental data set. Recalling the horizontal velocity to be large in the upper tropospheric jet region (which might be captured by the 250 hPa pressure surface), this result makes sense, as the jet region is better resolved in the operational than in the experimental data set and thus higher K_e and budget values are included in the vertical integration. By comparing the magnitudes for K_e and *div- ageo* at different levels, this result can be confirmed (Figure 5.2). Those two variables clearly reach maximum values in the upper troposphere, described by the 250 hPa pressure level, while their mid-tropospheric magnitudes are comparatively weak. The same result was found for the other budget terms also.

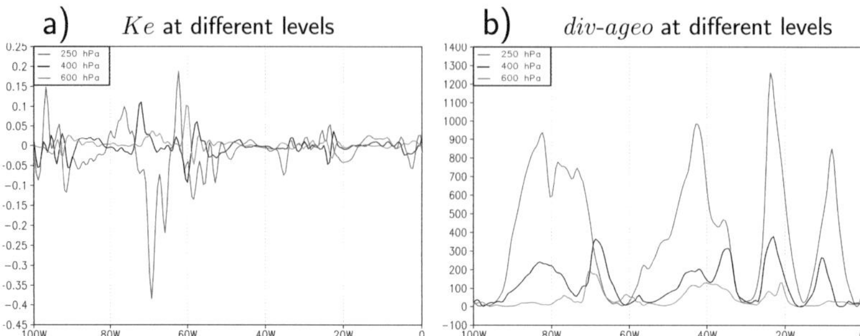

Figure 5.2: Magnitudes for a) K_e and b) *div-ageo* for lat=45°N at 7 September 2008, 00 UTC at 250 hPa (blue), 400 hPa (black) and 600 hPa(red) from the analysis data set.

Stronger differences occur for the baroclinic conversion term (*bcli*), whose computation is based on only six levels in the operational, while it can be computed at all ten levels in the experimental data set. Although *bcli* appears in the same regions in both data sets, it is weaker and thus only partly resembles the analysis values in the operational data set (Figure 5.1f). In the experimental setup, *bcli* matches the values that were found in the analysis data set (Figure 5.1e) more closely. The strongest differences are associated with the region of strong *bcli* to the northeast of the transitioning Hanna (red cross in both Figures). In the experimental resolution, *bcli* is a bit weaker, especially in the northern part of this maximum. Maximum values in the centre are 345 W/m^2 in the analysis and 324 W/m^2 in the experimental data set. In the data set with operational output resolution maximum *bcli* does not extent as far to the north than it does in the analysis (around 70-82°W and 42-49°N, Figure 5.1e). On the other hand, the values match quite well in the southern portion of the maximum. Baroclinic conversion in the operational resolution only reaches a maximum of 230 W/m^2 and thus differs about 100 W/m^2 from the maximum value in the other resolutions.

5.1.3 Generalisation and Selection of Data Set

The discussion above depicts only the differences between the individual budget terms at one specific time, i.e. the time close to the ET of Hurricane Hanna. However, the relationships can be confirmed by investigating the mean absolute error (MAE) for the simulated operational and experimental data set with respect to the absolute mean of the analysis data for a longer time period. The absolute mean and MAE are chosen as most of the quantities under investigation vary between positive and negative values, but their overall magnitudes are important. Furthermore, the MAE is chosen instead of the more common root mean square error (RMSE). As the RMSE is based on the square of the differences at each grid point, it puts more weight to large differences. In this comparison strong differences concerning the magnitude of the individual terms in the maximum region clearly exist in the different vertical resolutions. Thus, the MAE gives a better overview about the discrepancies that are caused by using data sets with coarser vertical resolution than the analysis, without penalising stronger deviations of the maximum values.

The MAE and absolute mean are computed in a region that captures most of the kinetic energy centres at investigation time, but neglecting centres in the polar region. The area reaches from 20°N to 65°N and from 110°W to 20°E and thus covers large parts of North America, the whole North Atlantic basin and the western part of Europe. The absolute mean of the analysis data set, together with the distribution of the MAE for both data sets show a common progression (Figure 5.3). As the results above already indicated, the operational resolution is closer to the analysis for all terms, except the baroclinic conversion, which is more sensitive to the coarse resolution in vertical velocity. The distribution of the MAE indicates that the mean of the individual budget terms at coarser output resolution follow their mean at the higher output resolution of the analysis data set, albeit their magnitude is smaller. Thereby, magnitudes of baroclinic conversion (*bcli*) and divergence of the ageostrophic geopotential flux (*div-ageo*) in the coarser data sets differ most strongly (with approx. half of the maximum value) from the magnitudes reached at high output resolution (Figure 5.3c,d, note differences in y-scaling). The weakest deviation between the coarse and high output resolution occur for the total generation term (Figure 5.3b).

For all terms except the baroclinic conversion, mean values of the quantities in operational output resolution are larger than in the experimental resolution, and thus closer to the values reached at high output resolution. Thereby, the magnitude of the differences for the generation term (Figure 5.3b), the convergence of the ageostrophic geopotential flux (Figure 5.3c) and the barotropic conversion (Figure 5.3f) between the two simulated data sets are less than the magnitude of the actual value in operational data set is. Convergence of kinetic energy flux with the total wind (Figure 5.3e) and the kinetic energy itself (K_e, Figure 5.3a), differ about the magnitude of the value, reached in operational output resolution. However, the mean baroclinic conversion in the operational resolution is clearly decreased (approx. 1.5 times), in comparison with the mean baroclinic conversion in the experimental resolution. Finally, the distribution of mean values, shown in Figure 5.3, indicates the magnitude of the averaged contribution of the individual terms to the budget of eddy kinetic energy. While the convergence terms (Table 3.1) contribute rather strongly (up to 25 W/m^2), the baroclinic, and especially the barotropic conversion have lower contributions (up to 10 or 6 W/m^2, respectively). This is reasonable, as the baroclinic conversion might be strong locally, but often occurs only in limited areas (regions of cyclone development). Thus the average over the whole domain may be smaller than the convergence terms, that are often spread over the whole wave pattern, and might be strong as well.

126

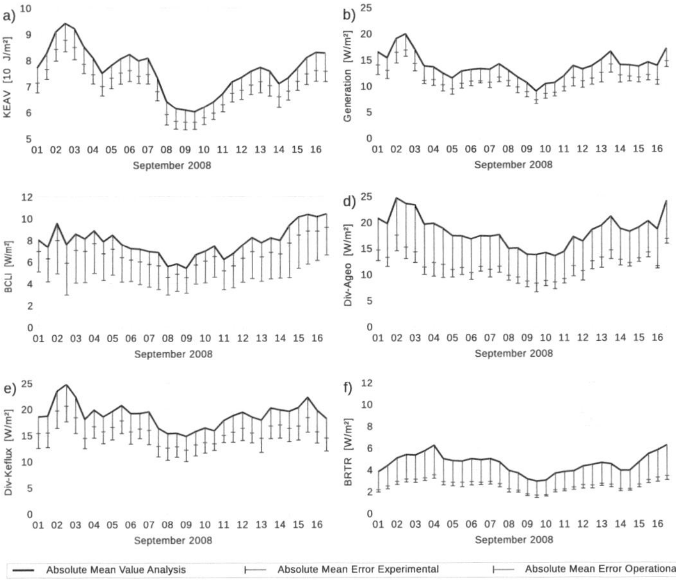

Figure 5.3: Absolute mean (black), MAE for the experimental (blue) and operational (red) data set for the budget terms, averaged over a region between 20°N to 65°N, 110°W to 20°E for each day in the first half of September 2008. As magnitude of quantities is lower, compared to the analysis values, only negative error bars are drawn. a) eddy kinetic energy (K_e), b) total generation (*generation*), c) baroclinic conversion (*bcli*), d) ageostrophic geopotential flux divergence (DLVP=*div-ageo*), e) divergence of K_e flux with total wind (DLKE=*div-keflux*), f) barotropic conversion (*brtr*). K_e in 10^5 J/m^2, budget terms in W/m^2.

The results above indicate that there clearly exists an impact of coarser vertical resolution on the terms of the eddy kinetic energy budget, as well as on the kinetic energy itself. This impact mainly affects the magnitudes of the values reached in the development during the considered time period. The spatial distributions of the several terms in coarser vertical output resolution often resemble the distributions in the higher resolved data set quite well. This can be confirmed, using the progression of the MAEs with respect to the mean values of the data set. For the baroclinic conversion, the operational vertical output resolution has larger errors than the experimental output resolution. However, the operational output resolution represents all other quantities to a better extent. Results indicate that the coarser vertical output resolution of both ensemble data sets should still allow a discussion of the eddy kinetic energy budget in different development scenarios, at least qualitatively. A quantitative conclusions on relevance of the individual budget terms among the different development scenarios and the real development can be drawn if the discussion stays within one data set with same vertical output resolutions and if the computation for the analysis budget considers only the pressure levels that are available for the ensemble forecasts as well. However, caution must be exercised when the quantitative contribution, i.e. from the transitioning storm on the mid-latitude flow, shall be compared with other studies. One further has to keep in mind that the magnitude of the baroclinic conversion might be too low, due to the coarse vertical output resolution.

In the remainder of this study, the experimental EPS is used for investigating the ET of Hurricane Hanna in 2008. The ET of Typhoon Choi-Wan in September 2009, will be examined in the operational ECMWF EPS, due to the lack of the experimental data set for this time. As the two case studies are carried out separately and mainly intra-case comparisons of different developments shall be drawn, the use of different data sets does not present any problems.

5.2 Application of EKE Analysis to Ensemble Forecasts

In this study the EKE analysis is applied to ensemble forecasts for ET events for the first time. ECMWF ensemble forecasts, initialised prior to an ET event, provide different possible scenarios for synoptic development during and after the transitioning process. A closer examination of the distinct development scenarios for the same transitioning storm using EKE analysis will then highlight the role of the transitioning tropical cyclone and the mid-latitude flow in the synoptic development.

Two ensemble forecasts, initialised prior to the ET of Hurricane Hanna in 2008 and of Typhoon Choi-Wan in 2009, respectively, serve as data base for this study. Both ensemble forecasts are characterised by a strong increase in standard deviation of the 500 hPa geopotential height after the storm was declared as extratropical system and thus are expected to show a broad variability of possible development scenarios to examine (Anwender et al., 2008) . First of all, the EOF- and cluster analysis presented in Chapter 3 is applied to the two ensemble forecasts in order to extract the main possible development scenarios. In addition to the well known application of the EOF-analysis to the 500 hPa geopotential height field, the EOF-analysis is now also applied to the vertically averaged eddy kinetic energy (K_e). Overall, this results in a number of clusters, depicting the main possible development scenarios that are found for the forecast under investigation. By comparing the EOF fields and clustering results for the geopotential height and the K_e clustering, differences between the two cluster analysis approaches are elaborated.

In most other studies (e.g. Harr et al., 2008, Anwender et al., 2008, Keller et al., 2011), the dominant development scenarios are then investigated using the cluster mean, i.e. the mean state which results from averaging over all members assigned to the cluster.. However, in this way detailed structures in the fields under investigation will be smoothed out due to averaging. Furthermore, the resulting mean state does not represent a consistent and physically developed solution for the atmospheric state, but rather an average over more or less distinct synoptic developments. These restrictions encouraged us to take a different approach here: One member of each cluster is defined to be representative for the synoptic development in this cluster. This representative member is the cluster member which has the closest Euclidean distance to the cluster centre. Then, this representative ensemble member is employed for further investigations.

The synoptic developments shown by the representative members of the geopotential height and the K_e clustering are compared to identify and select interesting ET scenarios within each forecast, i.e. a strongly reintensifying TC for one scenario and a dissipating TC as the contrasting scenario. Then the EKE-budget for selected cases is computed and studied in detail to diagnose the mechanisms that contribute to the different evolutions.

5.2.1 EOF- and Cluster Analysis of Different Quantities

In preparation for the detailed investigations of the different scenarios, the EOF- and cluster analysis is applied to the two forecasts to extract the main possible development scenarios. As discussed in Section 3.1.4, the geopotential height at 500 hPa is an appropriate quantity for the EOF analysis during ET events, as it represents features from the transitioning storm and characterises the mid-latitude flow pattern into which the TC is moving. However, this single level quantity only allows a two-dimensional representation of the atmospheric state at one specific height. The EKE analysis, applied to ensemble forecasts now affords the vertically averaged eddy kinetic energy as a new quantity for an EOF analysis. Due to the vertical average, K_e captures the atmospheric development at each available pressure level between 1000 and 100 hPa and provides a more detailed representation of the whole atmospheric state, seen in an energy framework. Furthermore, it is focused on regions with increased eddy kinetic energy, such as the transitioning storm itself or the flanks of the wave trains in the mid-latitudes. Through this analysis we expect to obtain a better separation of the development scenarios, depending on these "high-energetic" features.

The vertically averaged eddy kinetic energy is determined for each of the 51 ensemble members from the two forecasts. Then, as described in Section 3.1.4, the forecast time for the EOF- and Cluster analysis is selected using the standard deviation of the 500 hPa geopotential height. Thereafter, the EOF- and cluster analysis is applied to the K_e and to the 500 hPa geopotential height field as the specific forecast time. The EOF- and Clustering results are compared for a given forecast by visual comparison of the individual field and by using the Rand Index, as introduced in Section 3.1.5.

a) Results for Hurricane Hanna

Hurricane Hanna (Section 2.4.1) was a cat.1 hurricane that underwent ET in September 2008. A more detailed description of Hanna's life cycle is provided in Chapter 2.4. Due to its occurrence in the summer of 2008, some forecasts from the experimental ECMWF EPS are available for Hanna. The forecast that was initialised prior to the ET of Hanna and showed the strongest increase in 500 hPa geopotential height standard deviation after the transition of Hanna was chosen as forecast for further investigations. The selected forecast was initialised at 5 September 2008, 00 UTC and thus 30 hours before Hanna was declared to be an extratropical cyclone by the RMSC (7 September 2008, 06 UTC). A clear increase of uncertainty among the ensemble members occurs after the ET time of Hanna (Figure 5.4), downstream of the storm position between 60-0°W, propagating eastward. As in Chapter 4, the time for the EOF analysis (investigation time) is chosen to be the first forecast step after the strongest increase in standard deviation, which is 00 UTC, 9 September 2008 (Figure 5.4).

At the investigation time, the remnants of Hanna in the ensemble mean were located to the south of Greenland, interacting with a shortwave trough. A downstream trough developed to the west of Europe (Figure 5.5). Thus the EOFs are computed for the entire mid-latitude North Atlantic basin (0-80°W, 30-60°N).

EOFs for 500 hPa Geopotential Height

The strongest variability among the ensemble members of the forecast, as expressed by EOF 1 for 500 hPa, is clearly linked to the wave pattern in the mid-latitudes that is affected by the extratropical transition of Hanna (Figure 5.6). Overall, it resembles a weak amplitude pattern (Anwender et al., 2008) around the central Atlantic ridge, with strongest differences in the southwestern flank and in the amplitude of the southwards digging trough to the west of Europe. Weaker centres of action are found ahead (front flank) of the shortwave trough south of Greenland that interacts with Hanna, and in the vicinity of its rear (upstream) flank over New England, indicating a shift pattern (Anwender et al., 2008). About 36% of total variability, contained in the forecast is expressed by EOF 1. The second strongest variability among the ensemble members (EOF 2, Figure 5.6) is linked to the amplitude of the shortwave trough to the south of Greenland and to the flanks of the downstream trough to the west of Europe. EOF 2 explains about 20% of the total variability.

As in Chapter 4, the meaning of the identified EOF structures becomes obvious by comparing the 500 hPa geopotential height field of four members, contributing differently to these EOF patterns, with the geopotential height field of the ensemble mean (Figure 5.7, shading: members, contour: ensemble mean).

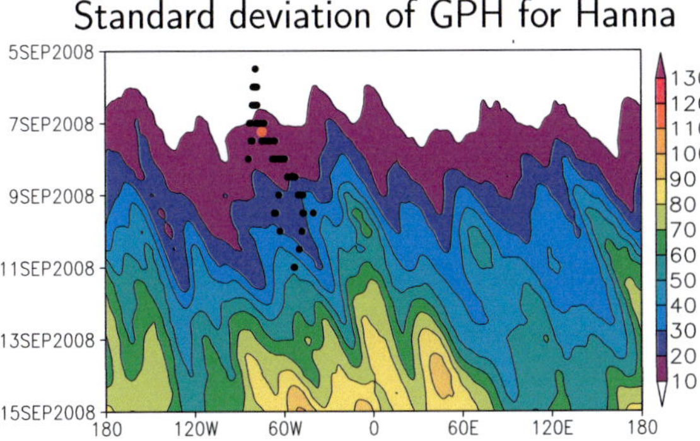

Figure 5.4: Hovmöller diagram for standard deviation of 500 hPa geopotential height (shaded, in *gpm*) in the ensemble forecast for Hanna, initialised 00 UTC, 5 September 2008. Black dots mark surface position of Hanna in the individual ensemble members, red dot marks its best track position as it was declared as extratropical system by the RMSC

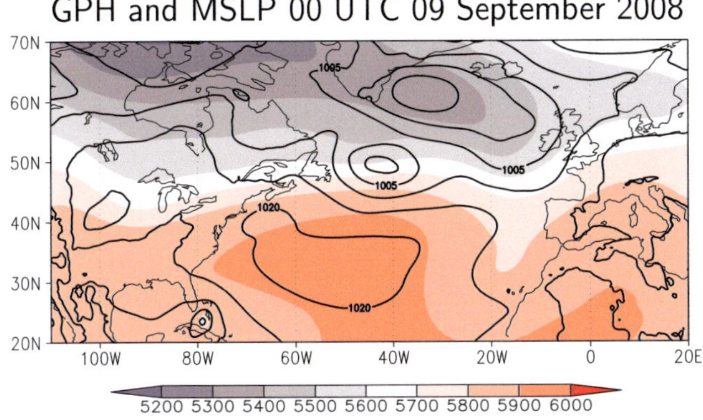

Figure 5.5: ECMWF ensemble mean of geopotential height at 500 hPa (in *gpm*) and mean sea level pressure (in 5 *hPa* increments) at investigation time for the EOF analysis of Hurricane Hanna (00 UTC, 09 September 2008)

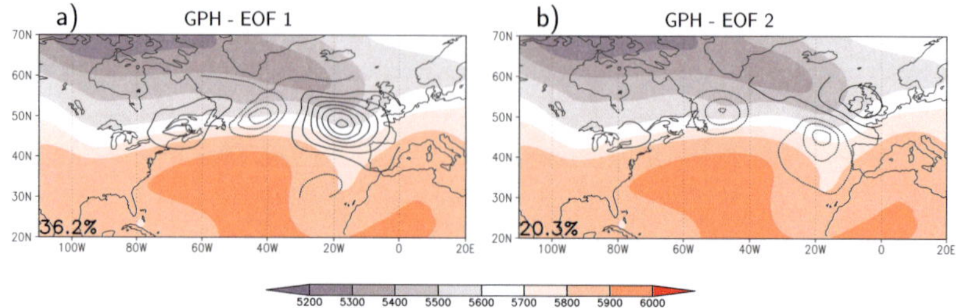

Figure 5.6: EOFs (in 10 *gpm* intervals, contours) for 500 hPa geopotential height (in *gpm*, shaded) for Hurricane Hanna at investigation time (00 UTC, 09 September 2008). a) EOF 1, b) EOF 2, captured amount of total variability at bottom left.

Members that contribute positively to EOF 1 are characterised by increased geopotential height to the west of Europe and over New England, but reduced geopotential height southeast of Greenland (Figure 5.7a). Thus, their downstream trough west of Europe is rather weakly amplified poleward of 45°N, which also results in a more zonal mid-latitude flow pattern over the North Atlantic. In contrast, the southern part of the trough (west of Spain and North Africa) is more pronounced. However, this member clearly indicates a shortwave trough already southeast of Greenland (axis near 40 °W) that is interacting with Ex-Hanna. In contrast, a rather strongly amplified downstream trough and a broad but flat central Atlantic ridge characterises members with negative contributions to EOF 1. Furthermore, some of these members do not show a shortwave trough to the south of Greenland (not shown), while others, like the given example, indicate this shortwave trough still southwest of Greenland (axis near 50°W, Figure 5.7b). Thus, beside the differences in amplification of the downstream trough, EOF 1 also separates members according to the position of the shortwave trough that is interacting with Hanna. A rather strongly amplified and southwards digging downstream trough, together with a moderately amplified central North Atlantic ridge and a clear shortwave trough (axis near 50 °W), interacting with Hanna are shown by members that contribute positively to EOF

134

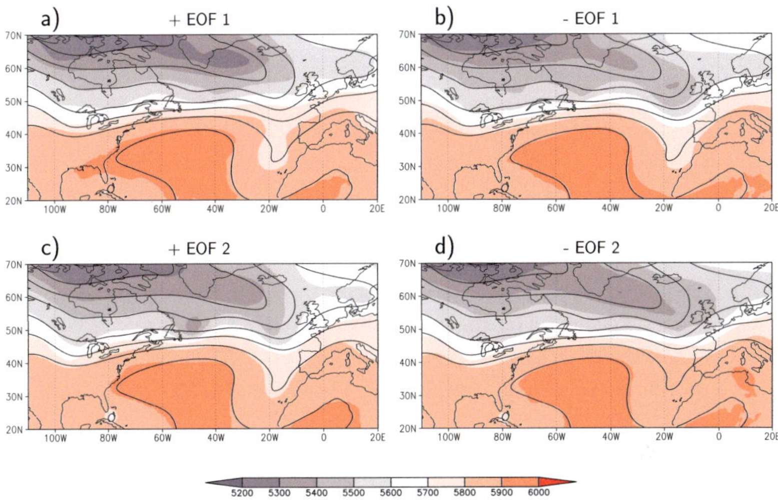

Figure 5.7: 500 hPa geopotential height (shaded, in *gpm*) and ensemble mean for 500 hPa geopotential height (contours, in *gpm*) for ensemble members with distinct contributions to the EOF patterns for Hurricane Hanna at investigation time (00 UTC, 09 September 2008). a) Member with positive contribution to EOF 1 and neutral to EOF 2, b) Member with negative contribution to EOF 1 and neutral to EOF 2, c) Member with positive contribution to EOF 2 and neutral to EOF 1, d) Member with negative contribution to EOF 2 and neutral to EOF 1.

2 (Figure 5.7c). Members with a negative contribution to EOF 2 have a rather flat mid-latitude flow, but a zonally aligned, thus northwards shifted, trough west of the UK (Figure 5.7d). Overall, geopotential height structure from members with distinct Principal Components (PCs) confirm the identification of amplitude- and shift pattern from Anwender (2007) in the EOF field of the forecast under investigation.

Fuzzy clustering (Chapter 3) of the ensemble members based on PC 1 and PC 2 leads to four different clusters with two more and two less well populated clusters. The clusters are separated clearly and also contain some extreme members (not shown). The cluster results are discussed later.

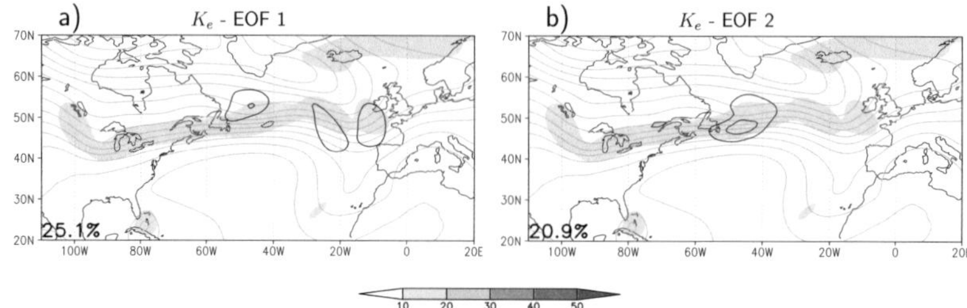

Figure 5.8: EOFs (in $3 \cdot 10^5 J/m^2$ intervals, black contours) for vertically integrated ensemble mean of eddy kinetic energy (in $10^5 J/m^2$, shaded) for Hurricane Hanna at investigation time (00 UTC, 9 September 2008), ensemble mean of 500 hPa geopotential height as grey contours. a) EOF 1, b) EOF 2.

EOFs for Vertically Averaged Eddy Kinetic Energy

After the vertically averaged eddy kinetic energy was determined for this forecast, the EOF analysis and the subsequent clustering procedure is applied to the K_e. As the K_e-distribution is often characterised by rather small scale disturbances and localised maxima, compared to the large scale 500 hPa geopotential height field, the EOFs for the K_e distribution are also expected to be more localised and smaller in size than they are for the 500 hPa geopotential height. Furthermore, high-energetic disturbances, such as a transitioning tropical cyclone should emerge clearly in the K_e distribution. The ensemble mean for the vertically averaged eddy kinetic energy (Figure 5.8) describes a long but rather zonal band of Ke between $40 - 60°$N, indicating the K_e distribution of the mid-latitude flow pattern.

Differences between the K_e distribution in the individual ensemble members and the ensemble mean mainly occur as dipole centres in the mid-latitude flow. Strongest uncertainty (EOF 1) is linked to the flow structure in the western North Atlantic and the circulation around the downstream trough to the west of Europe. About 25% of the total variability among the ensemble members occurs in these regions. Second strongest differences (EOF 2) clearly depict the vicinity of Ex-Hanna and thus capture distinct

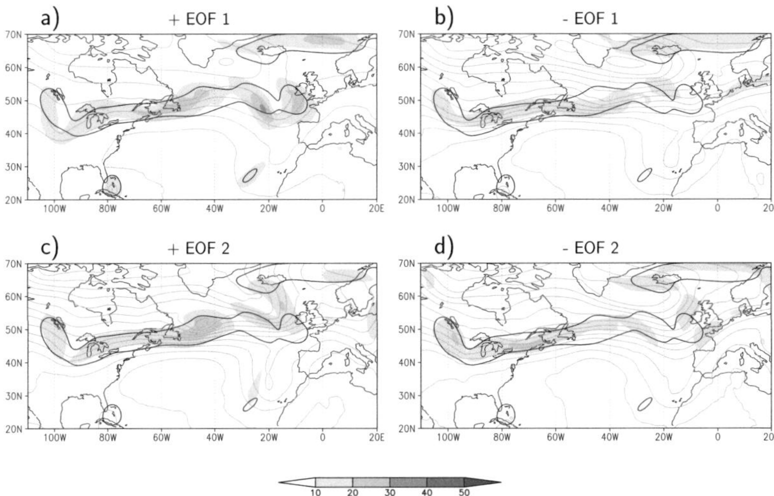

Figure 5.9: Vertically integrated K_e (shaded, in $10^5 J/m^2$) and ensemble mean for K_e (black contours, in $10^5 J/m^2$) for ensemble members with distinct contributions to the EOF patterns for Hurricane Hanna at investigation time (00 UTC, 09 September 2008), ensemble mean for 500 hPa geopotential height as grey contours. a) Member with positive contribution to EOF 1 and neutral to EOF 2, b) Member with negative contribution to EOF 1 and neutral to EOF 2, c) Member with positive contribution to EOF 2 and neutral to EOF 1, d) Member with negative contribution to EOF 2 and neutral to EOF 1.

representations of the transitioning storm and its circulation, embedded in the mid-latitude flow.

The dipole pattern in EOF 1 arises from the following differences in the flow field. A positive contribution to EOF 1 results from members whose K_e maximum over the central North Atlantic is shifted northward, resembling the 500 hPa ridge (Figure 5.9a). At the same time, their K_e maximum over New England and Newfoundland is shifted a bit southwards compared to the mean (in contours). The downstream trough west of Europe reaches more to the south than in the other members and coincides with a comparatively strong K_e centre at its rear- and a weaker centre on its front side. K_e structure in members with a negative contribution to EOF 1 is more zonally aligned without clear maxima over the North Atlantic. Especially to the south of Greenland, their K_e centre is to the south of that in the mean (Figure 5.9b). EOF 2 arises due to members with a strongly amplified shortwave trough and a coinciding K_e maximum to the south of Greenland (Figure 5.9c), and members which do not show the development of this shortwave structure at investigation time at all (Figure 5.9d). Furthermore, alignment of the K_e centre over New England differs with a weak ridge associated with a positive and weak trough with a negative contribution.

Clustering of the ensemble members, based on their PCs that resulted from the EOF analysis of the K_e distribution, also leads to four clusters that are well separated in PC1-PC2 phase space and do also capture some extreme cases.

Comparison of the Cluster Results

The two clustering approaches presented above resulted in two possible ways for how the 51 ensemble members can be grouped together, each consisting of four clusters and thus four distinct development scenarios. As the EOF patterns already indicated, differences among the ensemble members occur in slightly different regions, depending on which variable is used for the clustering. In the case of the geopotential height fields, the strongest differences were mainly linked to the downstream trough. Strong and dominant differences in the K_e distribution, however, also coincide with the K_e centre that is linked to the re-intensifying Hanna. This poses the question as to how the scenarios extracted from the geopotential height clustering differ from those scenarios that resulted from the K_e-clustering. This will indicate which field might result in a better separation of the members, depending on distinct flow features. Note that this comparison is only done for the synoptic development at investigation time (here 00 UTC 09 September

2008). At later times the developments are expected to differ noticeably, but should still be related to each other due to their similarities at investigation time.

First of all, the member assignments of the two cluster solutions are compared using the Rand Index, introduced in Section (3.1.5). This gives an indication of whether strong differences exist in the assignment at all, or whether the two clustering approaches mainly result in the same cluster solutions. With a Rand Index RI=0.75 the solution of the geopotential height (GPH) and the K_e clustering have some members assigned in the same clusters, but also some member that are assigned differently.

The contingency matrix allows the relation in cluster assignment between the GPH- and the K_e-clusters to be examined more closely (Table 5.2). It shows that each cluster has a group of "main"-members that are assigned together in the other cluster solution as well. For example, GPH-C3 and K_e-C1 have ten members in common, but also some members that are assigned differently (3 to K_e cluster 4). Strong resemblance is also found for GPH-C1 and K_e-C3. However, it is also obvious that some of the members are mixed up among the different cluster solutions. Furthermore, it is worth noting that the unassigned members change almost completely between the two cluster solutions, i.e. only one member is unassigned in both solutions (NoC from both solutions has only one member in common). From this we can conclude that the main development scenarios might be related somehow, but that details will change, depending on the members that are assigned differently. This outcome fits quite nicely with the basic idea behind the fuzzy clustering. Members are allowed to have a contribution to all clusters, and depending on the cluster centres, they might change clusters. Thus the pattern, which represents the cluster mean, might slightly shift from one scenario to another in the phase space of possible development scenarios.

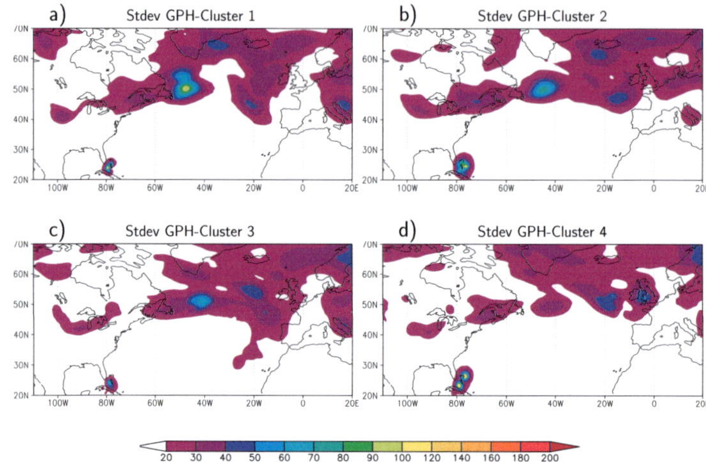

Figure 5.10: Standard deviation (in *gpm*) for 500 hPa geopotential height of the GPH-clusters for Hurricane Hanna at investigation time (00 UTC, 09 September 2008). a) GPH-cluster 1, b) GPH-cluster 2, c) GPH-cluster 3, d) GPH-cluster 4.

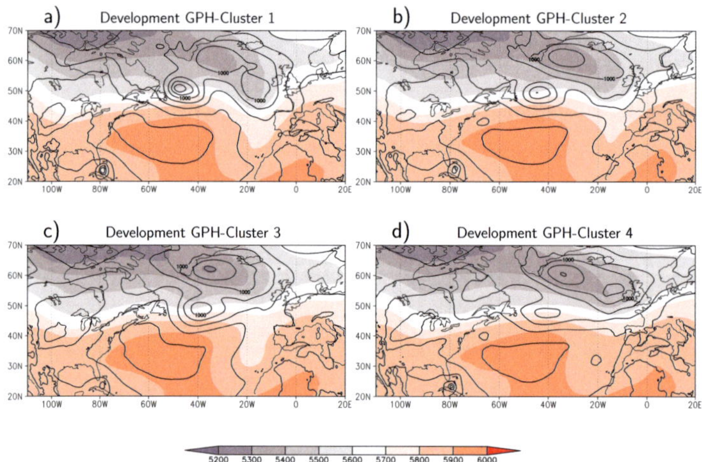

Figure 5.11: Cluster mean for 500 hPa geopotential height (in *gpm*) and mean sea level pressure (in *5hPa* increments) of the GPH-clusters for Hurricane Hanna at investigation time (00 UTC, 09 September 2008). a) GPH-cluster 1, b) GPH-cluster 2, c) GPH-cluster 3, d) GPH-cluster 4.

Table 5.2: Contingency matrix for the four K_e- and the four geopotential height (GPH)- clusters for Hanna. Numbers indicate, how many members the two clusters have in common, respectively. NoC stands for the number of members that are not assigned.

	K_e-clusters				
	C1	C2	C3	C4	NoC
C1	0	0	5	0	2
C2	1	7	1	4	1
C3	10	0	0	3	0
C4	1	3	0	0	4
NoC	4	2	0	2	1

(GPH-clusters label the rows)

The standard deviation of the members within one cluster from the related cluster mean now indicates how the assignment of different members influences the cluster mean and thus causes a possible shift towards another development scenario (Figure 5.10). Distribution of standard deviation furthermore highlights, where the separation of members due to the EOF- and cluster analysis fits well (i.e. where the members that are assigned together are closely related), and where it does not (i.e. development of members differ, although they are assigned together). By comparing the standard deviation patterns with the corresponding cluster means for the 500 hPa geopotential height field and the mean sea level pressure distribution, regions of increased uncertainty can clearly be linked to specific flow patterns. The strongest differences among the members within the four geopotential height clusters (Figure 5.10) occur south of Greenland, marked by a clear increase in standard deviation there. In this region a shortwave trough developed in some members and interacted with Hanna. This interaction is already represented differently in the distinct clusters (Figure 5.11). However, as the standard deviation shows, members in the cluster are not in agreement about the shortwave trough and its interaction with Hanna. Minor differences, strongest in cluster 4 (Figure 5.10d), are linked to the trough west of Europe. In contrast, members within the four

K_e clusters mainly differ in the representation of the downstream trough (Figure 5.13) to the west of Europe, where clearly enhanced standard deviation becomes obvious (Figure 5.12). However, differences among members in the K_e-clusters about the representation of the shortwave trough south of Greenland are rather minor. In all cases increased uncertainty occurs with respect to the development of Hurricane Ike, located in the Straits of Florida (80°W, 25°N).

By comparing the mean fields for the geopotential height clusters with the ones for the K_e clusters, we can identify a better representation and thus separation of distinct development scenarios for the downstream trough in the geopotential height clusters. They range from an only broad and shallow trough in GPH-cluster 4 (Figure 5.11d) to a rather strongly amplified and southwards digging trough in GPH-cluster 1 (Figure 5.11a). Both of these extremes occur in the K_e clusters to a weaker extent (Figure 5.13b,d). However, differences in the reintensification of Hanna and the strength of the shortwave trough south of Greenland are more marked in the K_e-clusters. A strongly reintensified Hanna, together with clear shortwave trough in K_e-cluster 3 (Figure 5.13c) juxtaposes a weak reintensification and no shortwave trough in K_e-cluster 1 (Figure 5.13a).

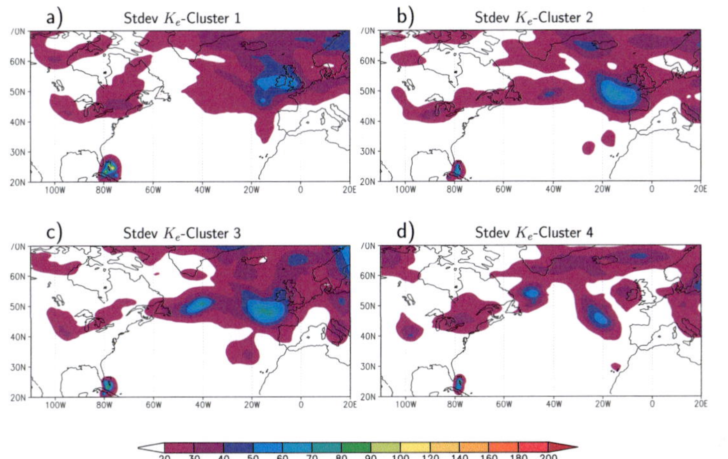

Figure 5.12: Standard deviation (in *gpm*) for 500 hPa geopotential height of the K_e-clusters for Hurricane Hanna at investigation time (00 UTC, 09 September 2008). a) K_e-cluster 1, b) K_e-cluster 2, c) K_e-cluster 3, d) K_e-cluster 4.

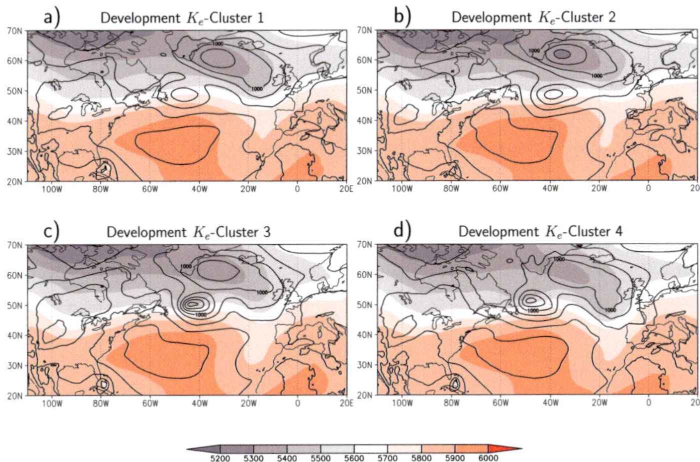

Figure 5.13: Cluster mean for 500 hPa geopotential height (in *gpm*) and mean sea level pressure (in *5hPa* increments) of the K_e-clusters for Hurricane Hanna at investigation time (00 UTC, 09 September 2008). a) K_e-cluster 1, b) K_e-cluster 2, c) K_e-cluster 3, d) K_e-cluster 4.

143

Summary and Discussion

In the case of Hurricane Hanna, the EOF analysis for the geopotential height and the vertically averaged eddy kinetic energy resulted in distinct EOF patterns, capturing different features of the flow field. EOFs for the geopotential height were mainly focused on the trough to the west of Europe, which lies downstream of the transitioning Hanna. They have relatively small amplitudes in the vicinity of a shortwave trough that interacts with Hanna. For the K_e field, the first EOF also focuses on the downstream trough, while EOF 2 gives prominence to the shortwave trough interacting with Hanna. The subsequent cluster analysis resulted in four clusters for both clustering approaches, which have some members in common, but also show some differences. These distinct assignments cause the development scenarios, stated by the cluster means, to vary slightly between the two cluster solutions. As it can be seen in the standard deviation of the members from the cluster mean, members that are assigned together in the geopotential height clustering coincide in representing the downstream trough, but show stronger differences in the vicinity of the shortwave trough close to Hanna. Members in the K_e clusters are more closely related in their representation of the shortwave trough, but differ more strongly in how they depict the downstream trough. Thus in this case, the K_e clustering is able to separate the processes in the vicinity of the reintensifying Hanna to a better extent, while it allows the members to be more different in the downstream trough west of Europe. In contrast, the geopotential height clustering better represents differences in the development of the downstream trough, while it allows distinct representations of the shortwave trough and the reintensifying Hanna. However, an overall agreement of main possible development scenarios of the two cluster is obvious.

b) Results for Typhoon Choi-Wan

Super-Typhoon Choi-Wan was the first cat.5 typhoon in the 2009 North Pacific season. As a strongly developed storm, Choi-Wan underwent ET in mid of September without making landfall. Choi-Wan's life cycle is described in detail in Section 2.4.6. Due to its occurrence in 2009, only operational ECMWF EPS forecasts are available for Choi-Wan. Based on the increase in standard deviation for the geopotential height field, one forecast is chosen for closer examination. The selected forecast was initialised on 00 UTC 15 September 2009 as Choi-Wan reached its peak intensity and thus well prior to its ET (12 UTC 20 September 2008). Although Choi-Wan was still in the tropics, nearly all ensemble members forecast the recurvature of Choi-Wan (Figure 5.14). However, the onset of recurvature and Choi-Wan's subsequent track, as well as its interaction with the mid-latitude flow was quite uncertain in the forecast under investigation. This is manifested in the very strong increase in standard deviation of geopotential height. Increasing uncertainty in the representation of a trough that will interact with Choi-Wan can already be identified after 00 UTC 17 September 2009 just east of 120°E (Figure 5.14). During its mature stage and its transitioning process, Choi-Wan approaches the mid-latitude trough in all 51 ensemble members. Just after Choi-Wan was declared as an extratropical cyclone a strong increase in standard deviation occurs in the vicinity of Choi-Wan itself and also affects the entire downstream region, at least towards the West Coast of North America (between 120-180°W). Thus, in the forecast under investigation the ET of Choi-Wan seems to strongly affect the mid-latitude flow pattern, causing a decrease in predictability for the downstream region including the North American continent.

The EOF- and cluster analysis is applied to the geopotential height- and vertically averaged eddy kinetic energy field on 00 UTC 21 September 2009. This is the first time after which the strongest increase in standard deviation is observed and after the storm was declared as an extratropical system (Figure 5.14). At this time, Choi-Wan was located east to east of Japan (Figure 5.15). As there was a maximum of K_e associated with the trough in the eastern North Pacific (140-180°W), the EOFs for the K_e are only computed for the western and central Northern Pacific, between 120°E-160°W and 30-60°N. For reasons of consistency and to compare the clustering results, EOFs for the geopotential height at 500 hPa are computed for the same area.

EOFs for 500 hPa Geopotential Height

Strongest differences among the ensemble members, captured by EOF 1, are linked to dominant features of the mid-latitude flow: the base and the rear flank of a shortwave trough south of Kamchatka, the front flank of a weak shortwave ridge just downstream between 165-180°E and the base of the shortwave trough east of Japan, which starts to interact with Choi-Wan in at least some ensemble members (Figure 5.16a). About 40% of the total variability among the ensemble members of the forecast in this region is expressed by EOF 1, which is a comparatively high amount. The second strongest variability only captures about 12% of the total variability and occurs mainly in the base of the shortwave trough near Kamchatka and towards its front flank and in the extension of the rear flank of the dominant trough over Alaska (Figure 5.16b).

Some members are characterised by a strongly amplified ridge east of Japan, reaching the Sea of Okhotsk between Kamchatka and Russia, an amplified and southwards digging trough around 180°E and a cut-off low due to Choi-Wan east of Asia (positive EOF 1 contribution, Figure 5.17a). Others have a clear trough over Kamchatka, expanding towards the east of Japan and a ridge in downstream regions around 180°E (negative EOF 1 contribution, Figure 5.17b). These contrasting synoptic developments cause the variability pattern identified by EOF 1. Differences captured by

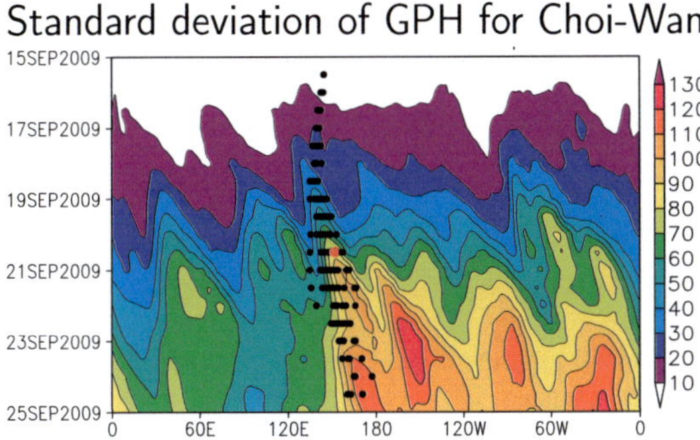

Figure 5.14: Hovmöller diagram for standard deviation of 500 hPa geopotential height (shaded, in *gpm*) in the ensemble forecast for Choi-Wan, initialised 00 UTC, 15 September 2009. Black dots mark surface position of Choi-Wan in the individual ensemble members, red dot marks its best track position as it was declared as extratropical system by the RMSC.

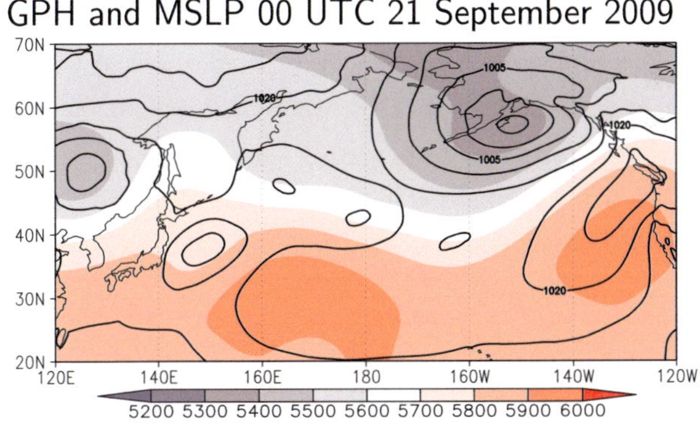

Figure 5.15: ECMWF-Ensemble mean of geopotential height at 500 hPa (in *gpm*) and mean sea level pressure (in 5 *hPa* increments) at investigation time for EOF analysis of Typhoon Choi-Wan (00 UTC, 21 September 2009).

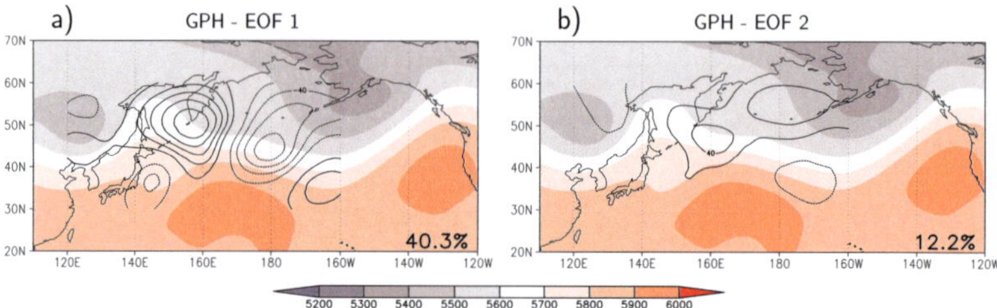

Figure 5.16: EOFs (in 10 *gpm* intervals, contours) for 500 hPa geopotential height (in *gpm*, shaded) for Typhoon Choi-Wan at investigation time (00 UTC, 21 September 2009). a) EOF 1, b) EOF 2, captured amount of total variability at bottom right.

EOF 2 are manifested in a pronounced secondary trough, with an associated anticyclonal wrap up of the ridge over Japan (negative EOF 2 contribution, Figure 5.17c), compared to an only weak secondary trough over the eastern North Pacific but a broad west- to central North Pacific ridge (positive EOF 2 contribution, Figure 5.17d).

The EOFs for the geopotential height at 500 hPa for Typhoon Choi-Wan are linked to the amplitudes, as well as to the flanks of the wave pattern. They differ from the findings from Anwender et al. (2008), as they do not only indicate a shift or an amplification of the whole wave pattern, but describe the zonal extension of the ridge. In positively contributing members the ridge is rather broad, while it is narrow in members with negative contribution to the EOF. In contrast to the strong variability associated with the ridge over Japan and the downstream wave pattern, the trough over eastern Russia is very similar in all four members. The EOFs partly capture this trough and indicate only weak variability. Hence, there exist a rather stationary flow structure with weak variability in the upstream region, directly beneath a strongly varying flow pattern in the western North Pacific, north and downstream of Choi-Wan. Fuzzy clustering of the principal components resulted in four well separated clusters with similar populations.

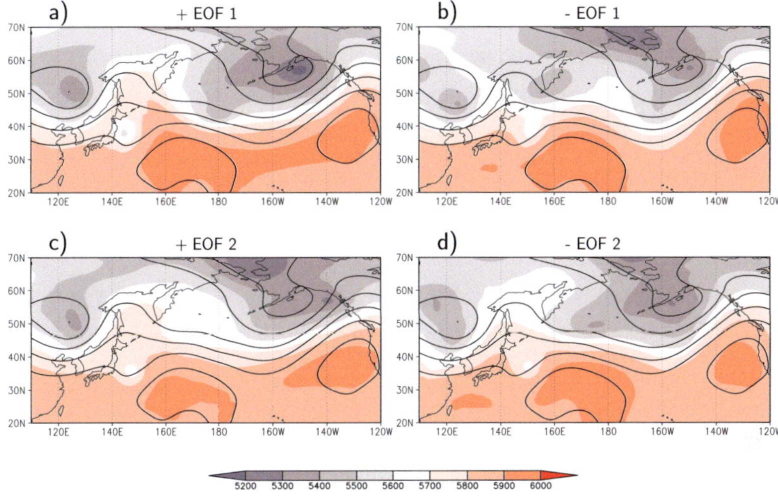

Figure 5.17: 500 hPa geopotential height (shaded, in *gpm*) and ensemble mean for 500 hPa geopotential height (contours, in *gpm*) for ensemble members with distinct contributions to the EOF patterns for Typhoon Choi-Wan at investigation time (00 UTC, 21 September 2009). a) Member with positive contribution to EOF 1 and neutral to EOF 2, b) Member with negative contribution to EOF 1 and neutral to EOF 2, c) Member with positive contribution to EOF 2 and neutral to EOF 1, d) Member with negative contribution to EOF 2 and neutral to EOF 1.

EOFs for Vertically Averaged Eddy Kinetic Energy

The representation of the vertically averaged eddy kinetic energy in the ensemble differs most strongly for the two K_e maxima associated with the confluent flow east of Kamchatka and with the rear flank of the central North Pacific trough, and to a lesser extent and with opposite sign in the maxima just east of Japan (Figure 5.18). With about 34% of the total variability, EOF 1 also captures a large amount of the uncertainty, contained in the forecast field at that time. The strongest uncertainty is linked to the northwards extent of the K_e maximum east of Japan, to the distribution of K_e in the trough over the Sea of Okhotsk and to the separation and southwards extent of the two K_e centres over the central North Pacific.

As it becomes clear from a comparison of members with positive and negative EOF 1 contributions, the variability captured by EOF 1 arises from the different representation of Choi-Wan, or, more specifically, different phases of the development (Figure 5.19a,b). In members with a positive EOF 1 contribution, (Figure 5.19a) Choi-Wan has reached the extratropics as an intense cyclone whose K_e centre already seems to support development of a strong downstream trough. In contrast, Choi-Wan is still located in the subtropics and no pronounced downstream development can be observed in those members that contribute negatively to EOF 1 and thus have stronger K_e in the negative EOF centre east of Choi-Wan (between 140-150°W and 32-42°N, Figure 5.19b). Members with a positive contribution to EOF 2 are characterised by a pronounced K_e maxima, which is elongated over the ridge in the Sea of Okhotsk, and their K_e centre east of Japan is located further north (Figure 5.19c). However, a negative contribution to EOF 2 is manifested in a weaker K_e centre to the north over Kamchatka, which is associated with a strong extratropical system south of Kamchatka. The K_e maximum east of Japan is located further south, its upper boundary is near 45°N (Figure 5.19d).

EOFs for the vertically averaged eddy kinetic energy field during the ET of Choi-Wan mainly capture the position and strength of the transitioning storm, its linkage to the downstream developing wave pattern and, to a lesser extent, the development of the ridge north of Choi-Wan. Fuzzy clustering of the PCs at investigation time was only stable for three clusters. However, as these three clusters partly spanned a large fraction of the PC1-PC2 phase space some alternatives are explored. The clustering procedure with four clusters resulted in two distinct solutions with similar occurrence. One of these solutions implied a boundary cluster with only two members, while the other clusters stayed nearly the same as for the three cluster solution. The other four cluster solution had equally populated and clearly separated clusters. As these four clusters pointed to distinct developments,

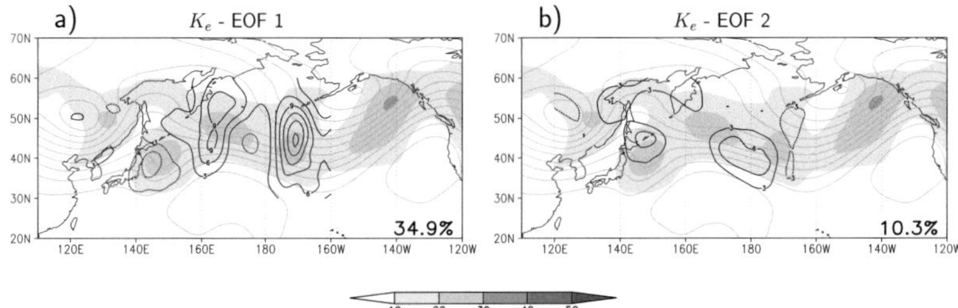

Figure 5.18: EOFs (in $3 \cdot 10^5 J/m^2$ intervals, black contours) for vertically integrated ensemble mean of eddy kinetic energy (in $10^5 J/m^2$, shaded) for Typhoon Choi-Wan at investigation time (00 UTC, 21 September 2009), ensemble mean for 500 hPa geopotential heights as grey contours. a) EOF 1, b) EOF 2, amount of total variability captured is given at bottom right.

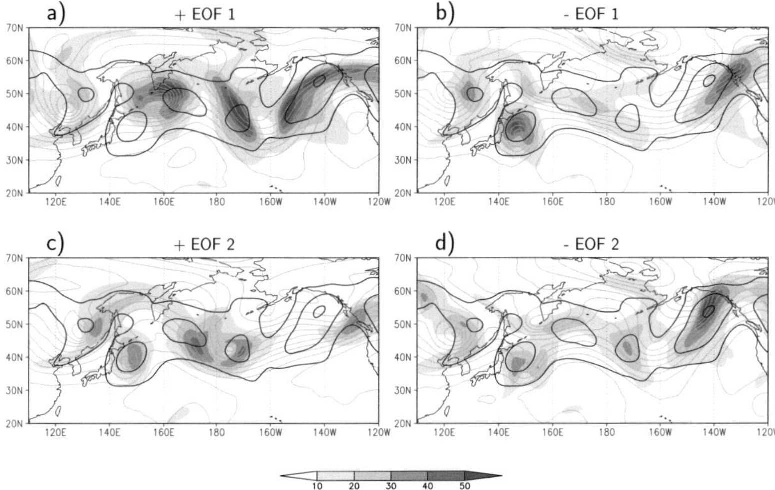

Figure 5.19: Vertically integrated K_e (shaded, in $10^5 J/m^2$) and ensemble mean for K_e (contours, in $10^5 J/m^2$, starting with 10 $10^5 J/m^2$, interval 10 $10^5 J/m^2$) with distinct contributions to the EOF patterns for Typhoon Choi-Wan at investigation time (00 UTC, 21 September 2009). a) Member with positive contribution to EOF 1 and neutral to EOF 2, b) Member with negative contribution to EOF 1 and neutral to EOF 2, c) Member with positive contribution to EOF 2 and neutral to EOF 1, d) Member with negative contribution to EOF 2 and neutral to EOF 1.

and as we are mainly interested in extracting different possible scenarios to compare physical processes, this solution was used for further investigations, despite it being a non-stable solution.

Comparison of the Clustering Result

A comparison of the two cluster solutions in terms of the Rand Index shows that the distinct cluster solutions for Choi-Wan with RI=0.74 are somewhat more strongly mixed than for Hanna. The contingency matrix (Table 5.3) indicates that the K_e cluster 2 and GPH cluster 4 are strongly related. The same applies to K_e-cluster 1 and GPH-cluster 1. The other K_e-clusters and GPH-clusters also have some members in common (GPH-C3 with K_e-C3 and GPH-C2 with K_e-C4), but contain a mixture from other members as well. As for Hanna, the different assignments also affect members in boundary regions which are assigned to a cluster in one, but are unassigned in the other solution. This result suggests that the captured development scenarios, as well as probably the variability among the ensemble members within one cluster might differ for the two clustering approaches. Surprisingly, although the cluster assignments indicated some mixtures among the members, the standard deviation patterns within the clusters for the geopotential height clustering resemble the corresponding

Table 5.3: Contingency matrix for the four K_e- and the four geopotential height (GPH)- clusters for Choi-Wan. Numbers indicate, how many members the two clusters have in common, respectively. NoC stands for the number of unassigned members.

		K_e-clusters				
		C1	C2	C3	C4	NoC
GPH-clusters	C1	7	0	0	4	0
	C2	0	0	2	6	1
	C3	1	0	5	3	5
	C4	0	5	2	0	3
	NoC	3	0	2	1	1

patterns for the K_e clustering to a good extent (Figures 5.20, 5.22a-d). In general, the standard deviation is stronger than in the case of Hanna, which is to be expected from the later investigation time (here +6 days, for Hanna +4 days). The extracted main development scenarios of the two clustering approaches do also coincide, although different representations of Choi-Wan and the environment mainly become obvious in the mean sea level pressure distribution (Figures 5.21, 5.23). Other than in the case of Hanna, where the standard deviation was maximised near the transitioning storm in the GPH-clusters and near the downstream trough in the K_e-clusters, standard deviation in the Choi-Wan case can not be linked to a different representation between the two cluster solutions. The corresponding clusters from the K_e- and GPH-clustering show related standard deviation patterns, that can be linked to distinct flow features.

A comparatively high amplitude trough over the western North Pacific and a stationary low in the Gulf of Alaska, whose secondary trough axis elongates southwestwards into the central Pacific, characterise the development in both cluster 1 at investigation time. Moderate remnants of Choi-Wan interact with a shortwave trough, embedded in a broad trough that is approaching from the Chinese main land. Strong outflow activity at upper levels comes with the remnants of Choi-Wan (not shown) and supports amplification of the ridge over Japan. Furthermore, cyclogenesis occurs ahead of the secondary trough axis over the central North Pacific (Figures 5.21a, 5.23a). Members in these clusters mainly differ in the representation of the shortwave trough, Choi-Wan is interacting with just off the Japanese East Coast (Figures 5.20a, 5.22a). As the position and intensity of the remnants of Choi-Wan vary, there is also variability in the downstream radiation of kinetic energy during the interaction between Choi-Wan and the mid-latitude flow.

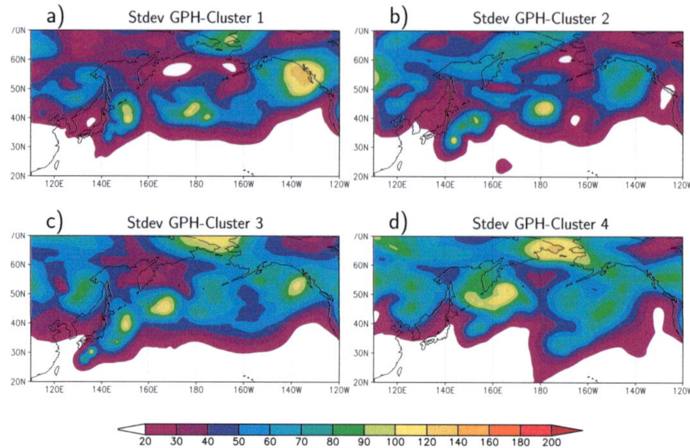

Figure 5.20: Standard deviation (in *gpm*) for 500 hPa geopotential height of the GPH-clusters for Typhoon Choi-Wan at investigation time (00 UTC, 21 September 2009). a) GPH-cluster 1, b) GPH-cluster 2, c) GPH-cluster 3, d) GPH-cluster 4.

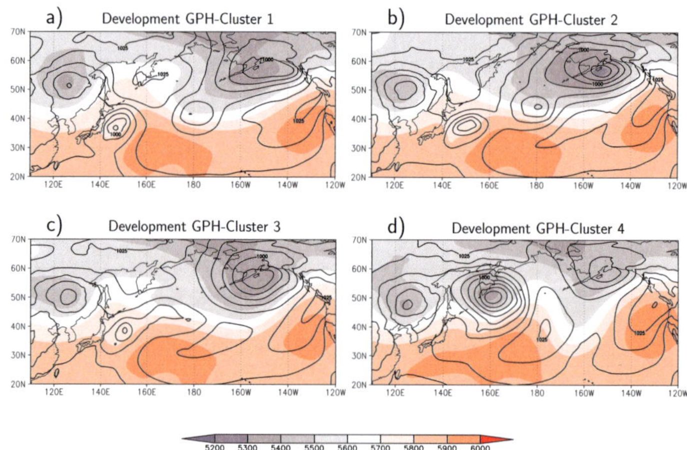

Figure 5.21: Cluster mean for 500 hPa geopotential height (in *gpm*) and mean sea level pressure (in *5hPa* increments) of the GPH-clusters for Typhoon Choi-Wan at investigation time (00 UTC, 21 September 2009). a) GPH-cluster 1, b) GPH-cluster 2, c) GPH-cluster 3, d) GPH-cluster 4.

In fact, this might then contribute to the increased uncertainty about the development of the secondary trough over the central North Pacific. Differences also occur in association with the stationary trough in the Gulf of Alaska. Overall, this pattern resembles Choi-Wan during its extratropical transition stage, as it is already interacting with the mid-latitude flow, i.e. causing further amplification of the ridge, but has underwent no strong reintensification or decay yet.

The mid-latitude wave pattern in the western and central Pacific is weaker in GPH-cluster 2 and K_e-cluster 4, respectively (Figures 5.21b, 5.23d). However, a pronounced ridge over the West Coast of North America and a strongly amplified stationary low over Alaska dominate the development. An associated secondary trough can be observed over the central North Pacific as well, but is not as strong as in clusters 1. The same is true for the cyclogenesis in front of it. Remnants of Choi-Wan are embedded in an open 500 hPa wave as well, but are located in vicinity of the ridge axis, instead of ahead of the trough approaching from the west. Strongest differences among the members are related to the position and elongation of the open wave, which is interacting with Choi-Wan (Figures 5.20b, 5.22d). Furthermore, distinct representations also occur in the amplitude of the secondary trough over the central North Pacific and the amplitude of the stationary low over Alaska. In this development, remnants of Choi-Wan seem to rather be in a subtropical phase. Although the transitioning storm is close to the mid-latitude flow, it has not started yet to influence the flow-pattern there.

A dominant trough over the Gulf of Alaska characterises the mid-latitude flow pattern in GPH- and K_e-cluster 3 (Figures 5.21c, 5.23c). Its secondary trough is small in the GPH-cluster 3 but more pronounced and associated with an extratropical cyclone in the K_e-cluster 3. Remnants of Choi-Wan are embedded in a slight wave in the vicinity of (GPH-cluster) or more ahead of the axis of a moderate, but narrow ridge over Japan. In this region,

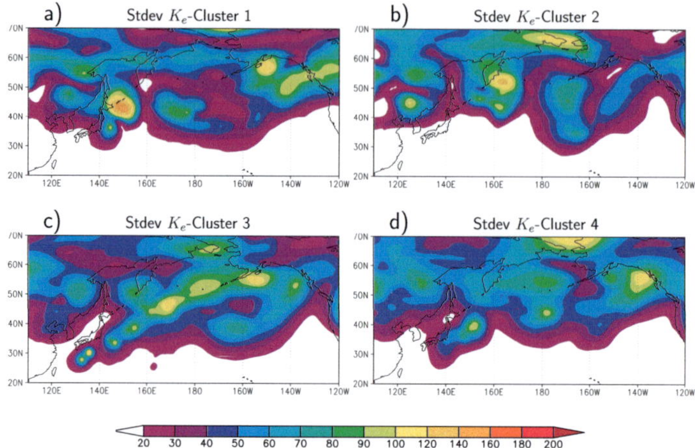

Figure 5.22: Standard deviation (in *gpm*) for 500 hPa geopotential height of the K_e-clusters for Typhoon Choi-Wan at investigation time (00 UTC, 21 September 2009). a) K_e-cluster 1, b) K_e-cluster 2, c) K_e-cluster 3, d) K_e-cluster 4.

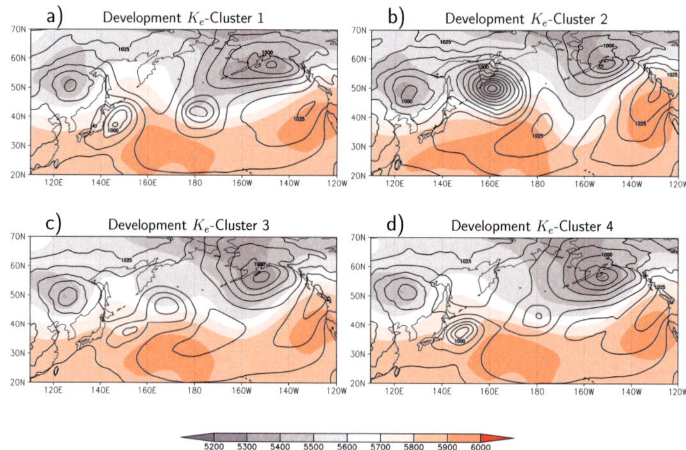

Figure 5.23: Cluster mean for 500 hPa geopotential height (in *gpm*) and mean sea level pressure (in *5hPa* increments) of the K_e-clusters for Typhoon Choi-Wan at investigation time (00 UTC, 21 September 2009). a) K_e-cluster 1, b) K_e-cluster 2, c) K_e-cluster 3, d) K_e-cluster 4.

the already weak remnants of Choi-Wan are prone to decay completely. Members in these clusters mainly differ in representing the wave structure associated with Choi-Wan, and presumably the amplitude of the secondary trough southeast of Kamchatka. Furthermore, they differ in the trough over the Behring Strait, which form the northern part of the stationary trough in the Gulf of Alaska (mainly GPH-cluster) and in representing the mid-latitude flow over Europe. It seems like the regions of increased uncertainty are linked to the occurrence of the distinct pressure systems. Taken as a whole, this development seems to resemble a non-reintensifying ET scenario.

The reintensification of Choi-Wan is underway in GPH-cluster 4 and K_e-cluster 2, respectively (Figs. 5.21d, 5.23b). A strongly amplified trough over Kamchatka favours the strong reintensification of the remnants of Choi-Wan. A weak ridge just downstream separates the Kamchatka-trough from another downstream trough, digging southwards from Alaska. However, the trough associated with Choi-Wan has already started to erode this downstream ridge. Variability among the members of these clusters is mainly associated with the position of the trough that Choi-Wan is interacting with. The strongly reintensified Choi-Wan might already influence downstream development by radiating kinetic energy through the adjacent ridge. Differences in the representation of this downstream radiation then enhances uncertainty in the representation of the strong downstream trough over the central Pacific. Differences also occur in the northern part of this downstream trough which can be identified over the Behring Strait. The strong reintensification character of this development is clearly obvious. Choi-Wan's transition seems to be almost finished, and the strongly reintensified extra-tropical cyclone starts to traverse the Pacific basin.

Summary and Discussion

The EOF analysis for the operational ECMWF EPS forecast, initialised prior to the ET of Choi-Wan, highlighted the strongest variability in the geopotential height field in the representation of the shortwave trough east of Japan and the eastward extension of a secondary trough over the eastern North Pacific. This variability patterns affected mainly the zonal extent of the ridge (broad or narrow ridge) over Japan and the trough over Kamchatka. For the K_e fields the strongest variability was linked to the downstream wave pattern in the central to eastern North Pacific, but also to the K_e maxima in the crest of the ridge over Japan. This variability results in clear differences in the amplification of the wave pattern over the central North Pacific and the ridge over Japan. Some uncertainty in the K_e fields was also found in the representation of the transitioning storm. Subsequent clustering then helped to extract four clusters for each approach. The main development scenarios as well as the distribution of standard deviation were related in the geopotential height and the K_e-clustering. These scenarios form a time line for the extra-tropical transition process. One scenario (GPH-cluster 2, K_e-cluster 4) has rather tropical characteristics, with Choi-Wan being in the centre of the ridge and not interacting with the mid-latitude flow. The set-up of the flow pattern resembles the northwest-pattern, as it was identified by Harr and Elsberry (2000). This development describes an early stage of the ET of Choi-Wan. Another scenario (GPH-cluster 3, K_e-cluster 3) has a related tropical set-up, but the mid-latitude flow structure into which Choi-Wan is moving is rather zonal and the remnants of Choi-Wan are already weak. Furthermore, the mid-latitude flow pattern has characteristics of the northeast-pattern (Harr and Elsberry, 2000). The dominant trough northeast of Choi-Wan in the Gulf of Alaska extents southwards a broad trough over the eastern to central North Pacific, and the wave pattern east of Choi-Wan is not further amplified. Thus, in this case Choi-Wan is prone to decay during the subsequent forecast hours. An advanced transitioning scenario is resembled by GPH- and K_e-

cluster 1. Moderate remnants of Choi-Wan have moved directly ahead of the northwestern trough and interact with the mid-latitude flow. Choi-Wan has already completed its ET and become a strongly amplified extratropical cyclone in GPH-cluster 4 and K_e-cluster 2, respectively. Furthermore, the ET of Choi-Wan has probably also influenced the downstream flow pattern. Ex-Choi-Wan is then moving across the Pacific.

From this point of view it might be argued that the members of the ensemble forecast under investigation mainly differ in their timing and might probably undergo related transitioning processes, but at different times, at least for three of the scenarios. This hypothesis is explored in the following section.

5.3 EKE-Budget for Selected Development Scenarios

The previous section highlighted the existence of a number of distinct possible development scenarios. In the case of Choi-Wan these scenarios spanned a particularly broad bandwidth of possible synoptic developments, caused by distinct positions of Choi-Wan with respect to the mid-latitude flow. Distinct development scenarios exist for Hurricane Hanna also. In the remainder of this chapter, a selection of possible development scenarios is examined for both storms. The EKE-Budget calculations are used to gain further insight into the underlying physical processes.

Until now, the development scenarios were depicted in terms of their cluster means (Figures 5.21). However, for the reasons of consistent developments and possible disadvantages due to smoothing, as discussed at the beginning of this Section (p.129), representative ensemble members are chosen for the subsequent analysis and discussion. These representative members for the different clusters are objectively selected by determining the Euclidean distance between each of the cluster members and the cluster centre in the

PC1-PC2 phase space (cf. Section 3.1.2, page 49). Then, we define this member to be the representative one, which lies closest to the cluster centre of the cluster under investigation. A comparison between the development scenarios, depicted by the representative members and by the cluster means shows how well the representative member describes the development, stated by the cluster mean. In addition to this more qualitative comparison, we can get a more quantitative measure for the similarity between the representative member and the cluster mean by computing the similarity index SI that was introduced in Chapter 3. A high SI between 0.7 and 1.0 is expected for those clusters whose members resemble a closely related development. If the development among the cluster members already differs, the representative member might only have a vague resemblance to the cluster mean, and the SI is expected to be below 0.7. However, although the representative member might differ from the cluster mean, it will be closer to the cluster mean than to the other scenarios, as we will see. Thus, it can still be seen as representative. Furthermore, the representative member mainly should match the development of the cluster mean prior to and around the investigation time. At later forecast times, differences among the members grow and cause the cluster mean to approach an even more smoothed development. The representative member might then not constitute a approximation of the cluster mean anymore, but is just one related scenario instead.

5.3.1 Forecast Scenarios for the ET of Hurricane Hanna

Each of the two clustering approaches for Hurricane Hanna resulted in four different development scenarios. Thus, a representative member has to be determined for each of the eight clusters. With a SI between 0.73-0.89 most of the representative members reproduce the cluster means quite well (Table 5.4). Only for K_e-cluster 2 the representative member apparently

Table 5.4: Similarity index between cluster mean and representative member for the forecast of Hurricane Hanna at clustering time. Upper three clusters show reintensification of Hanna, clusters in bottom row does not show a reintensification of Hanna.

Cluster	Similarity	Cluster	Similarity
GPH-c1	0.73	K_e-c2	0.23
GPH-c2	0.85	K_e-c3	0.77
GPH-c3	0.87	K_e-c4	0.89
GPH-c4	0.86	K_e-c1	0.85

differs more strongly from the cluster mean. These differences are associated with the representation of the mid-latitude wave pattern over Canada and Labrador, where the representative members has stronger amplified structures than the cluster members. The purpose of the study at hand is now to identify important processes during the ET of Hurricane Hanna that cause strong differences in the forecasts. Therefore, not all of the eight scenarios are investigated but rather contrasting ones. A closer examination of the development scenarios of the distinct representative members for Hanna showed that all of the development scenarios can be divided into two main groups. One of these groups is characterised by the interaction of the transitioning Hanna with a shortwave trough near 40-45°W, 50°N (Figures 5.11a,b,c and 5.13b,c,d), approaching from upstream. In this setup, Hanna underwent at least moderate reintensification over the central North Atlantic and caused a modification of the downstream flow pattern over Europe. In the other group of development scenarios (Figures 5.11d and 5.13a) no clear shortwave trough can be identified and the reintensification of Hanna to the south of Greenland is inhibited. This causes a distinct flow pattern over Europe. The group with reintensification scenarios more or less resemble the real observed development. One might ask, why the clustering led to four scenarios, while they can be separated in just two groups. Although the overall development can be divided into two groups, namely with and without reintensification of Hanna, the distinct development scenarios, stated by the cluster differ in detail. Thus, the phasing between Hanna and the trough might be different, which may cause distinct reintensification at different forecast times and also a variation in the downstream development. Two of the eight representative members (those for GPH-cluster 1 and 4) with strongest deviations in their synoptic development are chosen to be studied in detail. In the remainder, the representative member for GPH-cluster 4 is referred to as as scenario I, the member for GPH-cluster 1 as scenario II. At investigation time (9 September 2008, 00 UTC, forecast initialised at 5 September 2008, 00 UTC, cf. Section 5.2.1), members in scenario I are characterised by a rather zonal flow, with just a weakly amplified downstream trough west of Europe (Figure 5.24a, black isolines). Some members have a small shortwave trough south of Greenland, while the others do not or have their trough more to the west. The selected representative member (Figure 5.24a, red isolines) appears to be well centred in between the development scenarios, depicted by the other cluster members. A moderately pronounced shortwave trough south of Greenland in some members and a clearly amplified downstream trough west of Europe are found for scenario II (Figure 5.24b, black isolines). Here, the representative member also appears to be a good approximation of the development stated by the cluster, at least at investigation time and over the Atlantic basin, which is our region of interest (Figure 5.24b, red isolines).

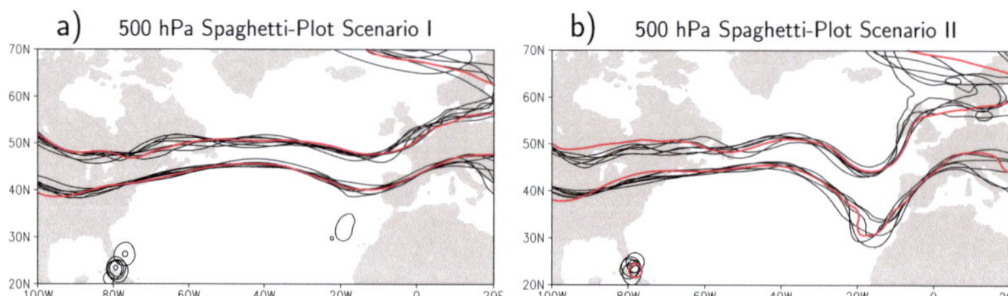

Figure 5.24: Spaghetti-Plot for 5600 and 5800 gpm height isolines at investigation time (9 September 2008, 00 UTC) for all ensemble members, assigned to GPH-Cluster 1 (a) and GPH-Cluster 4 (b) for Hanna.

a) Development Scenarios for Hanna and their Relation to the Analysis

Before starting the discussion about the ET of Hurricane Hanna in the EKE-framework, the two selected development scenarios are related to the analysed synoptic development during the forecast period. Such a comparison is facilitated by the anomaly correlation coefficient (ACC). It provides a good measure for correct forecasts of the spatial structure of the atmospheric fields under investigation and is not very sensitive to the magnitudes of the fields (Wilks, 1995). The ACC correlates the deviations (anomalies) of the predicted and the analysed field from a climatological distribution of the field under investigation

$$\text{ACC} = \frac{\sum_{m=1}^{M}\left[(y_m - C_m)(o_m - C_m)\right]}{\left[\sum_{m=1}^{M}(y_m - C_m)^2(o_m - C_m)^2\right]^{1/2}} \tag{5.1}$$

where y_m describes the forecast and o_m the observation at the m-th grid-point, C_m is the climatological value at this grid-point and results from an average over a fixed number of observations. If the forecast matches the analysis, the ACC is one, by increasing differences between forecast and observation it approaches zero. A forecast is referred to as useful and thus exhibit significant synoptic skill, until its ACC drops below 0.6 (Persson and Grazzini, 2007). Until the onset of reintensification in scenario II the two scenarios show closely related deviations from the analysis (Figure 5.25). After 12 UTC on 8 September 2008, scenario I starts to differ more strongly from the analysis as scenario II does and becomes insignificant in forecast skill after 00 UTC, 11 September 2008 (blue curve). However, scenario II continues to resemble the analysis quite well (red curve), as its correlation first stays near the general decrease in forecast accuracy in the deterministic ECWMF forecast (black curve). Around 10 September it starts to deviate more strongly from this reference to approach it again after a clear increase in the ACC around 00 UTC 13 September 2008. This member loses its forecast skill about 00 UTC 13 September 2008 and thus two days later than scenario I. Hence, scenario II were able to capture the real atmospheric processes during the ET of Hanna to a good extent, which is also manifested in the development of the Mediterranean cyclone due to the formation of the cut-off low (Grams et al., 2011).

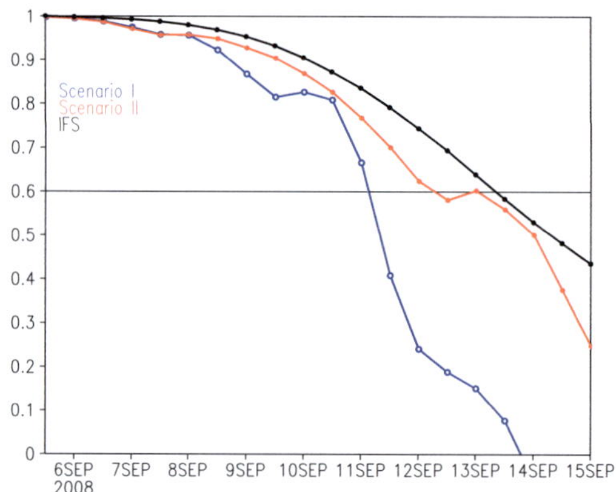

Figure 5.25: Anomaly Correlation Coefficient for the two members of the experimental ECMWF EPS forecast, the ECMWF analysis and the ERA-Interim climatology for the 500 hPa geopotential field between 30-70°N and 250-380°W for scenario I (blue) and scenario II (red). General forecast accuracy of the operational deterministic ECMWF forecast is highlighted in black. Forecast was initialised on 00 UTC, 5 September 2008. Thanks to Christian Grams for providing the ACC-Grads-Script.

b) Eddy Kinetic Energy Budget for Hanna-Scenarios

The two contrasting scenarios chosen as described above are used to examine the ET of Hurricane Hanna in more detail. These two development scenarios already start to differ 36 h after initialisation time, when Hanna is making landfall near the border between Florida and Georgia. Hanna's surface pressure centre is weak and the storm just clips the coast

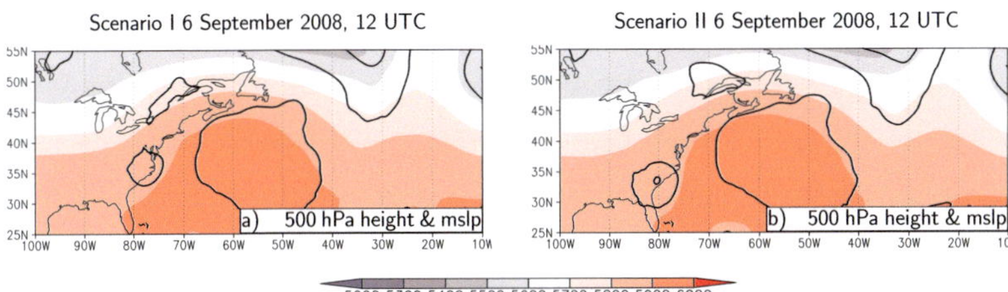

Figure 5.26: Scenario I (left) and II (right) 36 h after initialisation (6 September 2008, 12 UTC). Geopotential height (shaded, in gpm) and mslp (contours). Remnants of Hanna are embedded in the closed mslp contour near the US East Coast.

near Charleston to move quickly towards the northeast in scenario I. In scenario II Hanna moves further inland, has a stronger surface pressure centre and a slower northeastern velocity (Figure 5.26).

This lag in translation also affects the onset of interaction between Hanna and the extratropical circulation during the subsequent forecast hours. To consider differences in the ET itself, rather than differences at a specific forecast time, the developments of the two scenarios are shown at distinct times in the remainder of the discussion. The mid-latitude flow is characterised by a moderately amplified ridge just off the East Coast of North America, near Newfoundland, and a broad but weak trough over the main land (Figure 5.27a,b). A stationary low is located east of Greenland. The remnants of Hanna become embedded in a shortwave trough in the rear flank of the ridge near Long Island on 7 September 2008, 00 UTC in scenario I (Figure 5.27a). In scenario II, the remnants of Hanna can be found in a related position 12 h later (Figure 5.27b), but without a clear structure in the surface pressure and 500 hPa geopotential height field. However, the vorticity distribution at 850 hPa allows us to identify Hanna in scenario II south of Long Island, well off the coast (not shown). The time of the the official declaration of Hanna as an extratropical system (7 September 2008, 06 UTC) lies between the times of the scenarios shown in Figure 5.27.

Total fluxes of eddy kinetic energy (*totflux*) and net *generation* by advection of eddy geopotential heights already indicate that the interaction between Hanna and the mid-latitude upstream flow differs between the two scenarios (Figure 5.27c,d). In general, K_e is transported through the kinetic energy centre that is associated with increased wind speed in the broad ridge over Newfoundland (Figure 5.27c,d). In scenario I, the remnants of Hanna already support growth of this K_e centre, which is maximised northwest of Montreal, as *totflux* emanates the K_e centre of Hanna near 72°W, 42°N, towards the north and northwest. Hence, K_e fluxes originating from Hanna (causing net loss of K_e there) contribute to net *generation* of K_e between 70-80°W, 45°N (Figure 5.27c) and thus tend to accelerate flow through the ridge. Other regions of total net *generation* are found over and just off the coast of Labrador. Remnants of Hanna and the adjacent northeastwards extending baroclinic zone (not shown) act to baroclinically convert eddy available potential into eddy kinetic energy (Figure 5.27e) predominantly through the ascent of warm air masses (not shown). Most of this newly generated K_e is directly dispersed towards the K_e centre of the ridge by strong ageostrophic geopotential fluxes (*ageo*, Figure 5.27g), which diverge (*div-ageo*, negative) in the vicinity of Hanna and the baroclinic zone and converge northwest of Hanna, where the net *generation* area was identified in Figure 5.27c,(70-80°W, 45°N). Furthermore, *div-ageo*-convergence causes the net *generation* of K_e off the coast of Labrador (Figure 5.27c). The flux of K_e with the total wind (*keflux*, Figure 5.27i) acts to redistribute K_e from the upstream (western) to the downstream (eastern) portion of the K_e maximum. Due to the convergence region being at the rear flank of a downstream trough west of Europe, enhancement of the eastern portion of the K_e maximum may accelerate the flow at the rear of the trough, which will contribute to an amplification of the trough. Looking even further downstream, it becomes obvious that some energy fluxes originate in the centre of the ridge and support the northern K_e centre that is associated with the stationary low east of Greenland. Thus, in scenario I remnants of Hanna already start to act as an additional source of K_e during the ET, supporting the mid-latitude flow. As in the theory of downstream baroclinic development, this downstream dispersion is accomplished by diverging and converging ageostrophic geopotential fluxes.

In scenario II, there are a number of different features in the K_e budget. While the western North Atlantic ridge mainly resembles the pattern found in scenario I, the stationary low east of Greenland is more strongly amplified and a shortwave trough is identifiable to the southwest of the parent low. This causes a stronger height gradient east of Labrador and the K_e centre of the ridge reaches its maximum off the coast between 50-60°W, 53°N (Figure 5.27f), which is farther east than the maximum K_e in scenario I. At the same time, the K_e centre of the ridge is not as broad as in scenario I, especially between 40-50°W, but digs more to the southeast. Only a minor total flux of K_e exists between the remnants of Hanna and the mid-latitude K_e centre (Figure 5.27d), however, some K_e is generated over the Labrador peninsula (Figure 5.27d). Most of this generation is caused by baroclinic conversion, presumably due to ascending warm air masses in the northern part of the baroclinic zone over New England (Figure 5.27f). In contrast to scenario I, only minor baroclinic conversion is directly associated with the weak remnants of Hanna south of Long Island at this time. Diverging ageostrophic geopotential fluxes directly disperse the newly converted K_e over New England into the net *generation* regions over Labrador and Newfoundland (Figure 5.27h). Meanwhile, *keflux* redistributes K_e within the dominant centre from the rear to the front flank of the ridge and reaches regions that are farther south than in scenario I. This accelerates the flow towards the downstream trough and supports its further amplification. Downstream dispersion of K_e towards the stationary low east of Greenland via *div-ageo* is similar to scenario I. Overall, the impact of Hanna on the mid-latitude flow is weaker in scenario II than in scenario I at this time.

During the next 24 h (Figure 5.28), the remnants of Hanna continued to move northeastward. In scenario I the 500 hPa ridge over the western North Atlantic is flattened, due to the southwards digging broad but weak north-western trough over Labrador. The remnants of Hanna cannot be clearly identified from the surface pressure field (Figure 5.28a), but their associated vorticity maximum is located more to the south than in scenario II (not shown). Due to an approaching shortwave trough just upstream of the remnants of Hanna and the more southwards digging trough in the eastern North Atlantic, the western North Atlantic ridge is narrowed and a bit more amplified in scenario II (Figure 5.28b). This development is also associated with a more southwards digging downstream trough west of Europe (Figure 5.28a,b). While Hanna was stronger during the onset of ET in scenario I as examined with Figure 5.27, its remnants are now more amplified in scenario II and are located in a favourable position for further amplification.

The eddy kinetic energy budget now differs stronger between the two scenarios. While nearly all fluxes and *generation* of K_e are weakened in scenario I (Figure 5.28c), K_e is still radiated into the downstream trough in scenario II (Figure 5.28d). In scenario II, baroclinic conversion ahead of the remnants of Hanna clearly generates additional K_e (Figure 5.28f) in the eastern half of the storm. Strong baroclinic conversion ahead of Hanna

Figure 5.27: (Next page) Scenario I (left) and II (right) at the onset of interaction between Hanna and the mid-latitude flow. a, b) 500 hPa geopotential height at 500 hPa (shaded, in gpm) and mslp (contours, in 5 hPa increments); c, d) K_e (shaded, for all in $10^5 J/m^2$), *totflux* (arrows, for all in $10^6 W/m$, larger than $40 \cdot 10^6 W/m$) and *generation* (contours, for all in W/m^2); e, f) K_e (shaded) and *bcli* (contours); g, h) K_e (shaded), *ageo* (arrows, larger than $20 \cdot 10^6 W/m$) and *div-ageo* (contours); i, j) K_e (shaded), *keflux* (arrows, larger than $20 \cdot 10^6 W/m$) and *div-keflux* (contours). Contour interval $50 W/m^2$, starting with $50 W/m^2$ for all plots at each time, blue: divergence/loss of K_e, red: convergence/gain of K_e. See Table 3.1 for abbreviations.

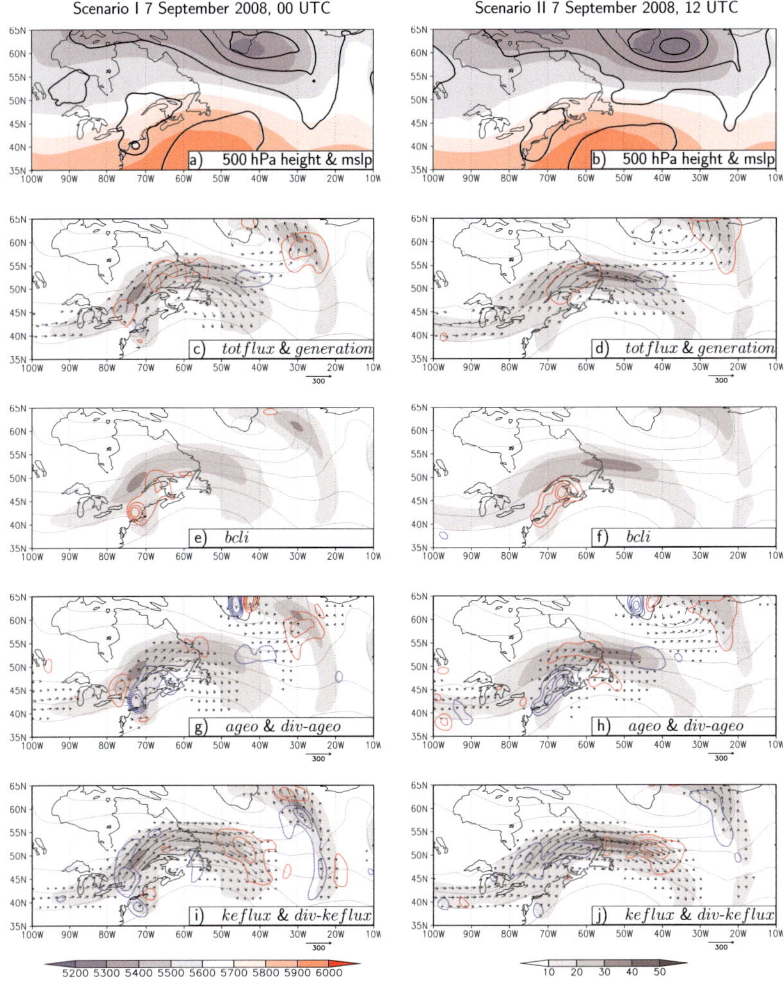
Scenario I 7 September 2008, 00 UTC
Scenario II 7 September 2008, 12 UTC
a)   500 hPa height & mslp
b)   500 hPa height & mslp
c)   totflux & generation
d)   totflux & generation
e)   bcli
f)   bcli
g)   ageo & div-ageo
h)   ageo & div-ageo
i)   keflux & div-keflux
j)   keflux & div-keflux
5200 5300 5400 5500 5600 5700 5800 5900 6000
10 20 30 40 50

was initiated (not show) as the system moved over New England towards its current position over Nova Scotia. But as most of this additional K_e is directly dispersed through the ridge towards the downstream trough via *div-ageo* (Figure 5.28h), net *generation* is rather weak in the vicinity of Hanna (Figure 5.28d). These fluxes, together with redistribution of K_e via *keflux* within the centre cause the downstream K_e centre at the rear of the downstream trough to increase and aid in the trough amplification. Thereby, the ageostrophic wind is enhanced in the base of the trough and enables further downstream dispersion of K_e towards the K_e centre in front of the downstream trough via divergent ageostrophic geopotential fluxes in the western and convergent ageostrophic geopotential fluxes in the eastern K_e centre. The differences in the two scenarios may result from the distinct amplification of the shortwave trough west of Hanna. The trough is more amplified in scenario II than in scenario I. As a result, the *keflux* from the rear to the front of the trough (Figure 5.28j) contributes to an increase in the flow and additional amplification of the trough. In turn, the trough amplification supports the reintensification of Ex-Hanna. Additionally, convergent ageostrophic geopotential fluxes in the very western end of the K_e centre over North America hint on slight energy support from upstream regions in scenario II. No clear flux and dispersion of K_e can be identified in scenario I. Minor baroclinic conversion and the lack of flux convergences in this scenario erodes the K_e centre associated with Hanna almost completely. At this stage, baroclinic conversion ahead of the transitioning storm, together with *keflux* in the vicinity of Hanna supports the onset of strong reintensification in scenario II and also supports the amplification of the downstream trough west of Europe in scenario II. Such a distribution of K_e, due to baroclinic conversion, and therefore the potential amplification of the downstream trough does not occur in scenario I.

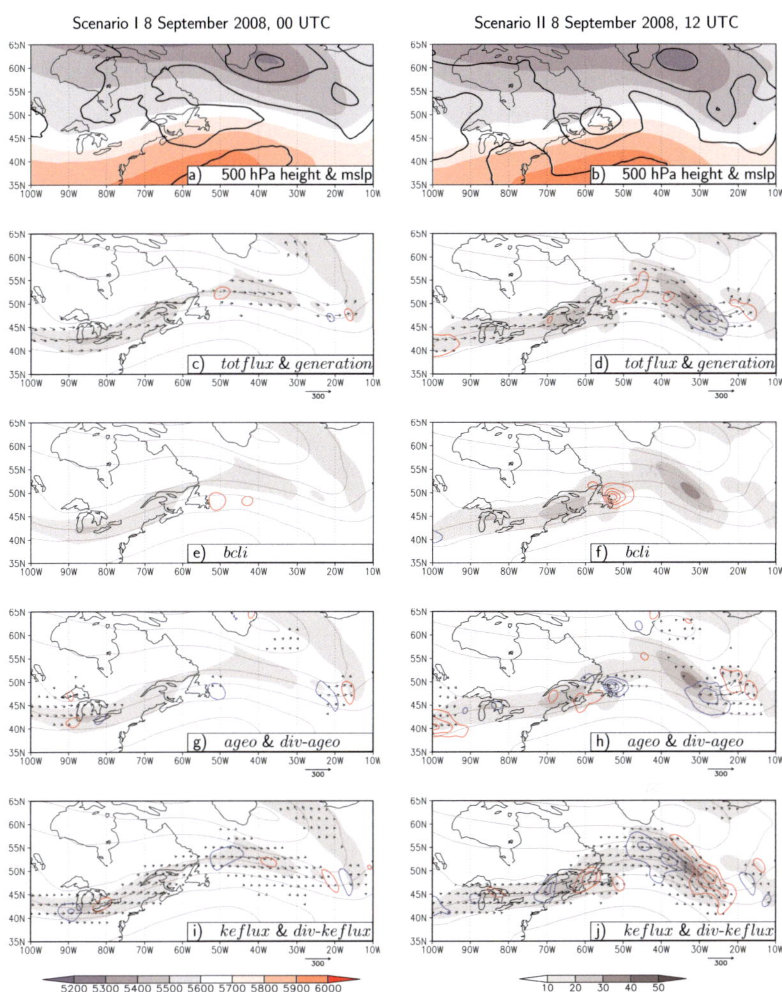

Figure 5.28: As Figure 5.27, but for the onset of reintensification in scenario II.

Due to its favourable position ahead of the shortwave trough, the remnants of Hanna undergo strong reintensification in scenario II during the subsequent 24 h (Figure 5.29). The merger of the shortwave trough that was west of Hanna with the Ex-Hanna circulation leads to an intensified cyclone that moved south of the primary low east of Greenland (Figure 5.29a). Energy fluxes start to recirculate K_e withing the centre of the cyclone that resulted from the reintensification of Hanna (Figure 5.29b). Ascending warm air ahead of the system and descending cold air to the rear (not shown) convert eddy available potential energy into K_e (Figure 5.29c). These regions are superimposed by diverging ageostrophic geopotential fluxes, which disperse the generated K_e towards the northern and southern flank of the K_e centre (Figure 5.29c). Hence, net *generation* of K_e occurs north and

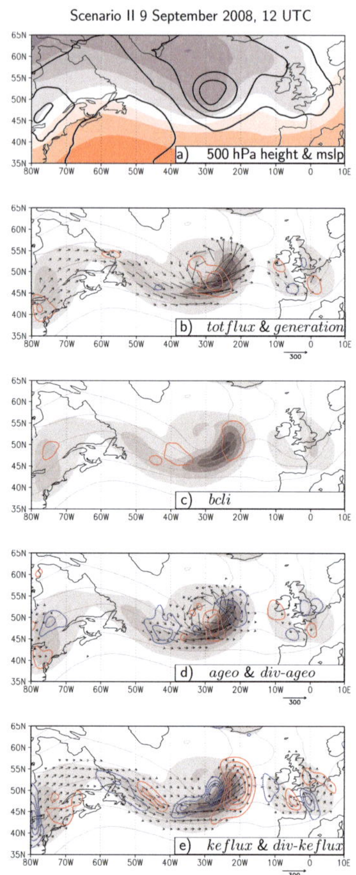

Figure 5.29: Scenario II during reintensification of Hanna. Plots as in Figure 5.27.

south of the reintensifying cyclone (Figure 5.29b). The flux of K_e act to recirculate K_e within the centre and accumulate K_e on its front flank (Figure 5.29e). The dominant recirculation of energy inhibits further downstream dispersion of K_e towards the trough west of the UK. As no net *generation* maintains this flow feature, the trough is prone to weaken and become a secondary part of the mid-latitude circulation during the subsequent forecast hours. At this time, no pronounced K_e centres or energy fluxes exist in scenario I. Hence, a detailed discussion for scenario I is unnecessary.

Another 24 h later (Figure 5.30) the reintensified merger in scenario II has circled around the dominant stationary low east of Greenland and supported its southeastwards expansion towards Europe. The system then forms a mature extratropical cyclone, positioned slightly ahead of a pronounced trough with southeastwards expansion from Greenland towards Europe (Figure 5.30 b, near 20°W, 52°N). A weak shortwave trough upstream of Hanna has formed in scenario I at this time as well, but between 20-30°N. The surface pressure of Ex-Hanna deepens only weakly, as the system is moving towards Ireland and the UK ahead of this disturbance. The stationary low east of Greenland is still the dominant feature of the mid-latitude flow in scenario I (Figure 5.30 a).

In scenario II, downstream dispersion of K_e from Hanna ceased to exist, the downstream trough attenuated into a weak cut-off cyclone near the Netherlands, that is nearly disconnected from the primary K_e centre associated with the reintensification of Hanna (Figure 5.30 d). Recirculating energy fluxes within the mature Ex-Hanna, on the other hand, accumulate to form a dominant K_e maximum in the southeastern quadrant of the cyclone. Ascending warm air in the tip of the K_e centre of Ex-Hanna, and descending cold air on the rear side of the cyclone convert available potential into eddy kinetic energy (Figure 5.30 f).

Ageostrophic fluxes (*div-ageo*) disperse K_e out of these regions, circulate it around the mature cyclone and finally converge southeast of the cyclone centre, between 20-30°W and 40-50°N (Figure 5.30 h), where a net *generation* of K_e is observed (Figure 5.30 d). This convergence of *div-ageo* in the tip of the trough enhances the geopotential gradient and thus accelerates the flow through the trough (Orlanski and Sheldon, 1995).

Fluxes of K_e with the total wind (*keflux*) also support recirculation, as they disperse K_e from the rear side of the maximum, associated with the *generation* region, to its front side (Figure 5.30 j), from where it is then re-circulated by ageostrophic geopotential fluxes again. Slight total K_e fluxes from the intensifying cyclone upstream near Montreal provide little energy to the Ex-Hanna circulation (Figure 5.30 d). Overall, the recirculating K_e fluxes appear to postpone decay of the mature cyclone, in accordance with Orlanski and Sheldon (1995) and to further amplify the mature trough near Europe during the next three days.

The development in scenario I is in strong contrast to this reintensification scenario. Ex-Hanna remains weak as it moves towards the northeast, and no pronounced *generation* or flux of K_e exists in the rather zonal flow over the whole Northern Atlantic basin. Recirculation of energy fluxes over Europe is missing, while *keflux* already disperse K_e further downstream towards Russia. Compared to the development in scenario II, this lack of energy generation, i.e. no baroclinic conversion, and continuous downstream dispersion appears to prevent reintensification of Ex-Hanna and subsequent amplification of the trough near Europe in scenario I.

At later forecast times, the observed differences in the ET of Hanna cause clearly distinct development scenarios for the Atlantic-European Sector. The geopotential height pattern remained rather zonal in scenario I, due to the absent K_e growth and recirculation. However, in the western North Atlantic, a strong and narrow trough develops, which also affects the flow pattern over Europe by downstream dispersion of K_e (not shown). Thus, on 14 September 2008, 00 UTC, the mid-latitude wave pattern in scenario I is characterised by a deep trough over the central North Atlantic, a slight ridge over the western Atlantic-European sector and a broad but weak trough over central Europe (Figure 5.31a), causing precipitation in states bordering the German Bay and the Baltic Sea (Figure 5.31c). Scenario II also develops an upstream trough which then erodes the dominant ridge over Europe (Fig-

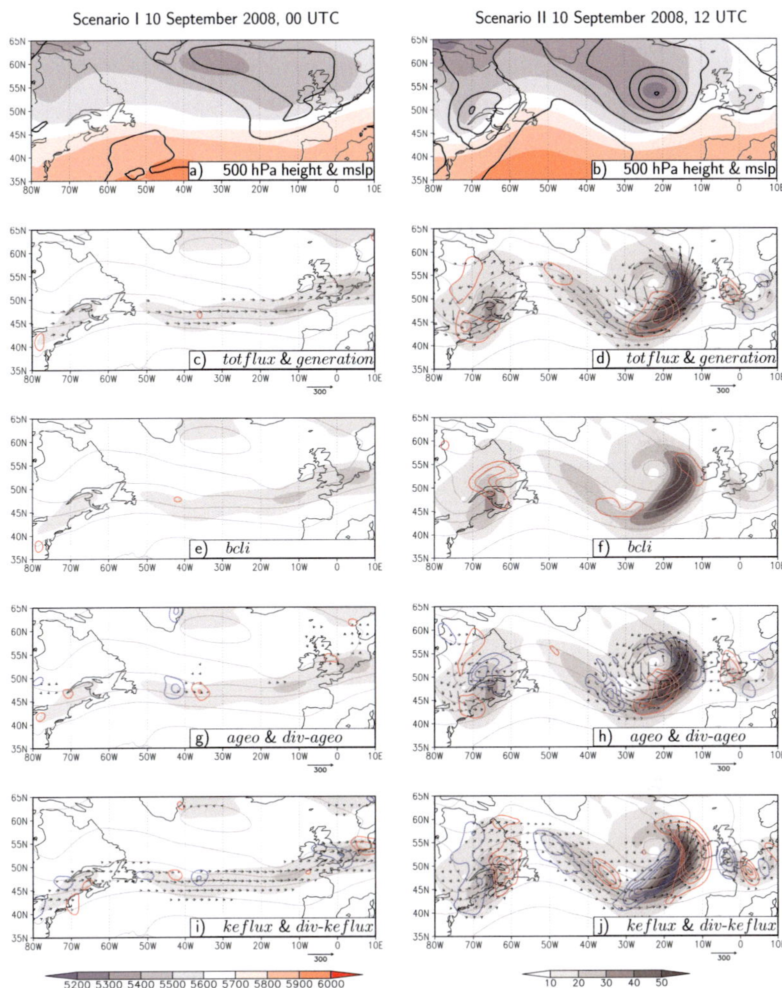

Figure 5.30: As Figure 5.27, but after completion of reintensification in scenario II.

ure 5.30) in the subsequent days. Due to this erosion, parts of the trough with Ex-Hanna and the associated K_e centre over western Europe start to cut off and propagate into the Ligurian Sea (Figure 5.31b). This causes the formation of a Mediterranean cyclone, associated with comparatively strong precipitation in the Provence and the Ligurian Sea (Figure 5.31d). Differences in the predicted development scenarios are also manifested in distinct 2m temperature fields, especially in the Alpine region (differences larger than 5 K), but also in northern parts of Spain, eastern Europe and in the whole UK (differences up to 5 K,Figure 5.31e,f).

c) Summary and Discussion for Hanna

Analysis of the EKE-Budget in the framework of downstream baroclinic development enabled us to interpret the influence of the transitioning Hurricane Hanna on the mid-latitude flow pattern. In scenario I, energy fluxes were strong during the earlier stages of ET, thus K_e propagated away from the transitioning storm towards the northwest and north into the mid-latitude flow pattern and further downstream into the stationary low near Greenland. This downstream support was fed by strong generation of K_e via baroclinic conversion in Ex-Hanna, while only minor baroclinic conversion was associated with the northeastwards extending baroclinic zone. Shortly after the remnants of Hanna moved off the North American main land a rather zonal flow without strong shortwave features was established over the Northern Atlantic basin. Baroclinic conversion and hence net *generation* of K_e collapsed and no advection of K_e emanated from the slight wave near 75°W and 48°N. During the next days the remnants of Hanna moved over the western Atlantic basin in this rather zonal flow. Between 20-30°W a shortwave disturbance developed directly upstream of Hanna and led to incipient reintensification of Ex-Hanna. The system propagates towards the UK but is not amplified further. Some days later a new trough

176

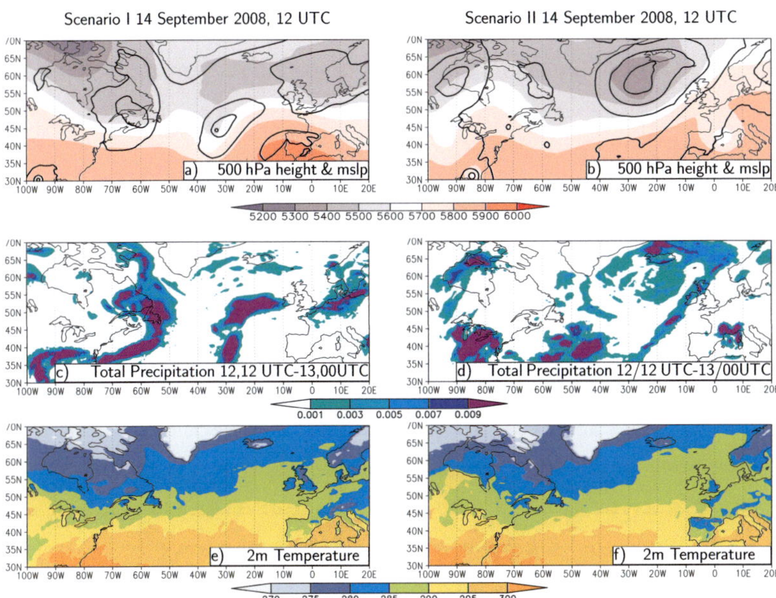

Figure 5.31: Distinct synoptic fields towards the end of the investigated forecast period for both development scenarios for Hurricane Hanna (valid at 14 September 2008, 00 UTC. a,b) 500 hPa geopotential height (shaded, in gpm) and mslp (contours, in 5 hPa increments) for scenario I and II, c,d) sum of large scale and convective precipitation (in m) between 13 September 2008, 12 UTC and 14 September 2008, 00 UTC for scenario I and II, e,f) 2 m Temperature (in K) for scenario I and II.

disturbance in the upstream region helped to slightly amplify the wave pattern over Europe, via downstream radiation of K_e.

In contrast, energy fluxes emanating from the transitioning Hanna and K_e *generation* via *bcli* were rather weak during the onset of ET in scenario II. On the other hand, the northwards extending baroclinic zone had a stronger contribution to net K_e *generation* via baroclinic conversion. However, the K_e generation region was superimposed by diverging ageostrophic geopotential flux that dispersed K_e into the K_e centre of the ridge. This predominantly north- and northeastwards directed downstream K_e dispersion weakened as the remnants of Hanna propagated towards the North Atlantic and a recirculation of K_e around Ex-Hanna was triggered. As Hanna coupled with a shortwave disturbance, approaching from Canada, baroclinic conversion intensified and continued generation of K_e. Recirculation of K_e prevented the K_e centre of Hanna from decaying by reason of strong downstream dispersion, but supported amplification of the associated trough near Hanna instead. Although downstream dispersion of K_e in the earlier development stages supported amplification of a downstream trough, this structure could not reach a mature stage as K_e fluxes from Hanna ceased to exist, because no innate baroclinic conversion maintained the centre. Due to reintensification and minor downstream dispersion, the trough associated with Hanna then became the dominant mid-latitude flow feature west of Europe. In the subsequent days, this trough preconditioned the development of a Mediterranean cyclone, triggered by an approaching upstream trough.

In summary, differences in the baroclinic conversion of K_e and downstream dispersion via ageostrophic geopotential fluxes during the early ET stages of Hanna apparently influence the formation of the two distinct development scenarios for the downstream regions. In addition to differences in the storm motion between scenario I and II, there also exist differences in the onset of strong baroclinic generation of K_e in the remnants of Hanna. While strong baroclinic conversion associated with Ex-Hanna has already begun as the cyclone is located near Long Island in scenario I, strong baroclinic conversion near Ex-Hanna in scenario II does not occur until the remnants of the cyclone propagated over New England and reached Nova Scotia (cf. Figure 5.27e,f). During the early onset of *bcli* in scenario I, downstream dispersion via ageostrophic geopotential fluxes from Ex-Hanna towards the K_e centre of the ridge is strong (Figure 5.27g). Hence, most of the newly generated K_e near Ex-Hanna is dispersed directly towards downstream (and upstream) regions in scenario I. Then, *bcli* nearly ceases to exist before Ex-Hanna moves off the mainland (Figure 5.28e). Only a minor shortwave disturbance is located directly upstream of Ex-Hanna. This disturbance is not amplified further, presumably due to a lack in generation and advection of K_e. Due to only weak energy fluxes from upstream and no recirculation in the mid-latitude flow pattern Ex-Hanna starts to decay and a zonal flow is established over the Atlantic basin. In contrast, downstream dispersion of K_e via *div-ageo* is slightly decreased already during the later onset of strong *bcli* in scenario II (not shown). In addition, *bcli* in scenario II is sustained until Ex-Hanna abandons the coast and moves ahead of the short-wave trough. The interplay between advection of K_e around this shortwave trough (Figure 5.28j), baroclinic conversion in Ex-Hanna (Figure 5.28f) and the onset of recirculating energy fluxes supports the further amplification of the trough and in turn the reintensification of Ex-Hanna (Figure 5.29a). Thus, the outlast of baroclinic conversion in Ex-Hanna, together with re-circulating K_e fluxes and only minor downstream dispersion is apparently crucial for amplification of the mid-latitude wave pattern after the ET of Hanna. Although dispersion of K_e towards the downstream trough during ET and before the onset of strong reintensification preconditioned the development of a downstream trough west of Europe, a strong trough there is not established until the mid-latitude trough, associated with Ex-Hanna approaches Europe.

The important role of baroclinic conversion of available potential into eddy kinetic energy in the formation of distinct development scenarios during the ET of Hanna coincide with the findings of Grams et al. (2011). By applying trajectory calculations and a potential vorticity (PV) perspective[3] (Hoskins et al., 1985), along with potential vorticity inversion (Davis and Emanuel, 1991, Grams, 2011) on ECMWF analyses, they investigate the crucial role of diabatic processes during the ET of Hanna. They identify the cross-isentropic advection of air masses with low PV in the warm conveyor belt, associated with Hanna's upper-level outflow to support ampli-

---

[3]Potential vorticity is conserved during adiabatic processes, and thus net generation of PV directly hints on the existence of diabatic processes or advection

fication of the downstream ridge. Another warm conveyor belt emanating from the new upstream cyclone then later helps to thin the trough associated with Hanna, which finally leads to the formation of a PV streamer and the onset of a Mediterranean cyclogenesis. Furthermore, latent heat release due to condensational processes leads to generation of new PV in the vicinity of Hanna, as warm and moist air masses in front of the cyclone are advected towards the baroclinic zone and become lifted. These vertical motions are also associated with the generation of K_e due to baroclinic conversion. However, the relationship between the findings of Grams et al. (2011) and the investigation of the ET of Hanna using the EKE analysis could be proven further, if the EKE budget for the analysis data would be taken into account, instead of the budget for developments from the ensemble forecast. Already in the early approach by Palmén (1958), baroclinic conversion of available potential into kinetic energy was identified as an important source for K_e during the ET process. The importance was further highlighted by the findings of Harr and Dea (2009). In their examination of the ET of TY Nabi and TY Banyan, they also identified the crucial role of baroclinic conversion of K_e due to rising warm air predominantly in the eastern half of the transitioning cyclone, in accordance with the conceptual model of Klein et al. (2000). In the framework of downstream baroclinic development, this additional generation of K_e due to the transitioning storm sustained the upstream K_e centre and supported the strong amplification of the downstream wave pattern, which clearly indicates the possible importance of a transitioning tropical cyclone for the amplification of the downstream wave pattern. In the case of Hanna, no strong downstream development occurs, if the baroclinic conversion of K_e ceases to exist during the ET process. If baroclinic conversion near Hanna is maintained through the ET, it first facilitates downstream development in the central North Atlantic and presumably supports the amplification of the shortwave trough near Hanna. As dispersion of K_e from Hanna towards the downstream wave pattern ceases to exist, also does the amplification of the downstream wave pattern.

5.3.2 Forecast Scenarios for the ET of Typhoon Choi-Wan

The EOF- and cluster analysis of the operational ECMWF EPS forecast for Typhoon Choi-Wan (initialised 15 September 2009, 00 UTC) also resulted in four distinct clusters for both clustering approaches. Based on the Euclidean distances, representative members are defined for each of the overall eight clusters. Thereby, two of the GPH-clusters and two of the K_e-clusters have their representative members in common. This fits well with the findings that the eight clusters describe four main development scenarios (Section 5.2.1). The similarity index for the 500 hPa geopotential height field of the cluster mean and the representative members (Table 5.5) is less than it was for the case of Hurricane Hanna. However, this is reasonable because of the different forecast times. In the case of Hanna, the clustering procedure was applied on day four of the forecast, while the forecast for Choi-Wan is investigated on day six. Thus, due to the later forecast time the individual ensemble members may differ more strongly than they do for an earlier forecast time. A strong increase in differences among the ensemble members becomes obvious also in the Hovmöller diagram for the standard deviation of the 500 hPa geopotential height field (Figure 5.14). GPH-cluster 4 and K_e-cluster 2 have the same representative member as do GPH- and K_e-cluster 3. Thereby, the SI between the former ones is highest in this case (see Table 5.5) with $SI = 0.79$ and $SI = 0.82$, respectively. However, the representative member for cluster 3 for the GPH and K_e clustering has

Table 5.5: Clusters and similarity index SI between cluster mean and representative member for the forecast of Typhoon Choi-Wan. Related cluster are in same line.

Cluster	Similarity	Cluster	Similarity
GPH-c1	0.64	K_e-c1	0.67
GPH-c2	0.78	K_e-c4	0.54
GPH-c3	0.44	K_e-c1	0.39
GPH-c4	0.79	K_e-c2	0.82

less similarity to the associated cluster mean ($SI = 0.44$ and $SI = 0.39$, respectively). The distribution of the SI also indicates that members in cluster 3 are more different in their development than the members in GPH-cluster 4 and K_e-cluster 2. Recalling the associated development scenarios, discussed above (Figs. 5.21 and 5.23) we discover that rather good agreement occurs between the cluster mean and the representative members for the scenario with a strongly reintensified Choi-Wan, towards the end of its ET (GPH-C4, K_e-C2). The rather weak resemblance between the cluster means and the representative members is associated with the decaying scenario (GPH- and K_e-cluster 3). For each of the four main development scenarios, one representative members is chosen for further investigations, those are the ones for GPH-clusters 2,3[4],4[5] and K_e-cluster 1. The latter is chosen, as it has a higher SI than the representative members for its corresponding GPH-cluster.

Due to the longer forecast time, the geopotential height patterns among the members differ more strongly than in the case of Hanna. However, all members that are assigned to the distinct clusters agree in their overall scenario (Figure 5.32), while there exist clear differences in between the several clusters. The K_e-cluster 1 is representative for the onset of Choi-Wan's reintensification (cf. Section 5.2.1). All of its members are characterised by a broad, but flattened ridge reaching from Japan to Kamchatka, a broad downstream trough and a clear shortwave trough that interacts with Choi-Wan. The representative member describes a middle ground, especially in the representation of the broad ridge and downstream trough pattern, while it has a rather strong amplification of the shortwave trough-ridge pattern near Japan (Figure 5.32a). Stronger differences with regard to the amplification of the northern ridge and the position of the shortwave trough linked to Choi-Wan exist between the members of the GPH-cluster 2 (Figure 5.32b). However, these members clearly coincide in their representation

---

[4]or K_e-cluster 3
[5]or K_e-cluster 2

of the development of the broad, but weak amplified downstream trough. The representative member is also found to resemble a middle path of development. There is more uncertainty in the geopotential height cluster that describes the non-developing case in Section 5.2.1 (GPH-cluster 3). Its members clearly differ in the amplification and structure of the ridge over Japan, in the position and amplitude of the shortwave trough as well as in the development directly downstream (Figure 5.32c). Members in the cluster that showed the strong reintensification of Choi-Wan (GPH-cluster 4), concur in having a pronounced and narrow downstream trough near 160°W, and a rather moderate ridge over Japan, which coincides with the upstream region of the reintensified Choi-Wan. Some members of this cluster also have a shortwave trough east of Honshu (150°E, 35°N). The representative member is clearly embedded in the centre of the development in the downstream trough, but shows a strongly amplified and cyclonically breaking ridge at around 170°E in this cluster. This is a slight aberration from the development in the other cluster members, however one other member shows a related scenario as well, but with formation of a cut-off low over eastern Siberia. As the representative members overall coincide quite well with their associated cluster means, the representative members should capture the timeline in the ET development as well (Section 5.2.1, p. 158). The distinct development scenarios characterised during examination of the clustering are now allocated as follows: the scenario which is still in its tropical phase is referred to as scenario I (K_e-cluster 1), scenario II refers to the decaying one (GPH-cluster 2), scenario III has already completed its ET and shows a strongly reintensified extratropical cyclone (GPH-cluster 4) and in scenario IV, the ET of Choi-Wan is still under way (GPH-cluster 3).

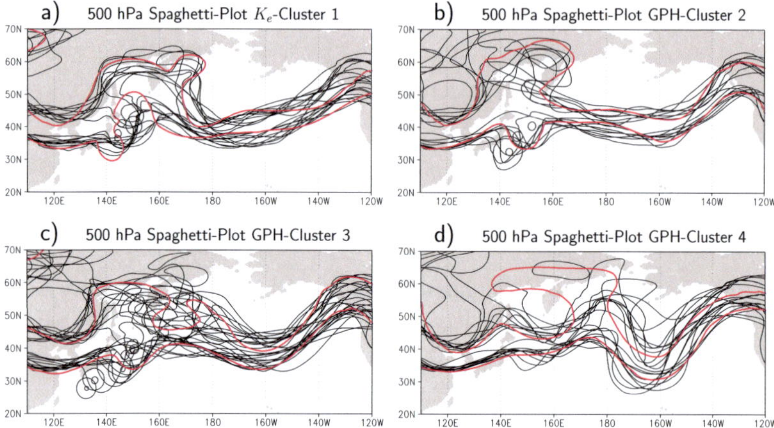

Figure 5.32: Spaghetti-Plot for 5600 and 5800 gpm height isolines at investigation time for all ensemble members, assigned to K_e-Cluster 1 (a), GPH-Cluster 2 (b), GPH-Cluster 3 (c) and GPH-Cluster 4 (d) for Choi-Wan.

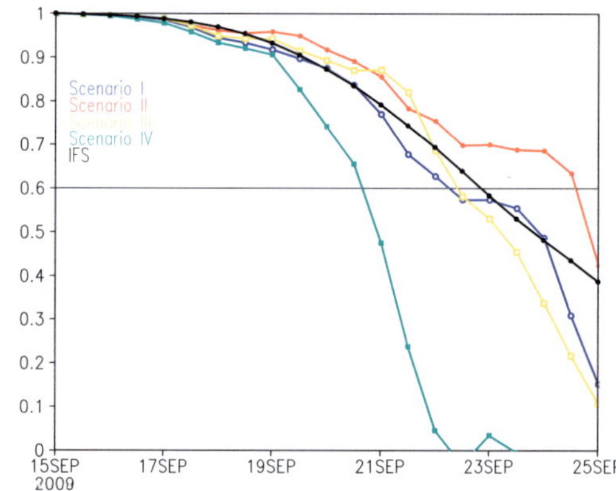

Figure 5.33: Anomaly Correlation Coefficient for the four members of the operational ECMWF EPS forecast for Typhoon Choi-Wan, the ECMWF analysis and the ERA-Interim climatology for the 500 hPa geopotential field between 20-60°N and 120-300°W for scenario I (blue), scenario II (red), scenario III (yellow) and scenario IV (green). General forecast accuracy of the operational deterministic ECMWF forecast is highlighted in black. Forecast was initialised on 00 UTC, 15 September 2009.

a) Development Scenarios for Choi-Wan and their Relation to the Analysis

The different synoptic developments, examined for Choi-Wan, are now also compared with the verifying ECMWF analysis, using the Anomaly Correlation Coefficient (ACC). Two of the four development scenarios capture the real synoptic development of the 500 hPa geopotential height field quite well (Figure 5.33). Namely, those are the members, described by scenario II (red) and III (yellow). Especially, on the medium range, they capture the real atmospheric development better than the deterministic forecast in average does on that forecast time (black curve). The merger of Ex-Choi-Wan and the extratropical cyclone is apparently better described by scenario III, as confirmed by a direct visual comparison of the 500 hPa and surface pressure fields (not shown). However, the cut-off process in scenario III then clearly diverges from the analysis, causing a drop in the ACC around 22 September 2009. The development in scenario I (blue) is not that far away from the analysis, but is not as good at capturing the real analysis, as the other two are. Finally, the strongly amplified ridge north of Choi-Wan in scenario IV is apparently far away from the real development, as its ACC soon looses its synoptic significance (drop below 0.6) and is far away from the other developments after four forecast days.

b) Eddy Kinetic Energy Budget for Choi-Wan-Scenarios

A closer examination of the selected development scenarios for Choi-Wan, using the EKE analysis will now help to identify the role of the transitioning Choi-Wan in the modification of the mid-latitude flow pattern. The forecast under investigation was initialised well before the onset of ET. During the first few forecast days the development of Choi-Wan and the mid-latitude flow pattern coincided quite well in all of the representative members. However, the scenarios start to diverge around four days after the initialisation time. Thereby, strongest differences are associated with the amplification of the mid-latitude wave pattern pattern and the location of Choi-Wan relative to this pattern. As in the case of Hanna the scenarios are compared at different stages of the ET of Choi-Wan. The stages for investigation are chosen by a comparison of the synoptic developments in the distinct scenarios. The forecast times at which the ET process is in a related stage in the distinct scenarios are then taken into account in the following discussion. Hence, the forecast time between the different scenarios vary, but the scenarios depict related stages in the ET of Choi-Wan and highlight differences in the transformation process.

The transitioning typhoon has no clear influence on the extratropical flow, manifested in K_e-fluxes into the mid-latitudes, until its remnants couple with a shortwave trough east of Japan. At this time, 00 UTC 20 September 2009, the strength and position of the surface pressure centre of Choi-Wan varies just slightly between the scenarios, with closely related positions in scenario I-III (Figures 5.34a,b, 5.35a), but with a more southeastwards position in scenario IV (Figure 5.35b). However, the position of Choi-Wan relative to the mid-latitude flow pattern, as well as the strength and position of an extratropical cyclonic disturbance northeast of Choi-Wan differs clearly between the scenarios. The strength of the mid-latitude cyclone, together with its position are somewhat related in scenario II and III, but in scenario III, the southeastwards digging trough northeast of Japan (140-160°E, 45°N) is stronger, which is also associated with a more amplified downstream ridge (Figure 5.35a). Due to the stronger trough, Choi-Wan is located in a favourable position for reintensification ahead of the trough in scenario III. At the same time, the geopotential gradient between the trough and the subsequent downstream ridge just east of Choi-Wan is enhanced and the northeastward flow around the eastern flank of the transitioning storm is strengthened. In the other scenarios, the remnants of Choi-Wan become embedded in the ridge, located upstream. Scenario I has a more strongly amplified extratropical cyclone at a more western position. The interaction between Choi-Wan and the mid-latitude flow seems to be delayed in scenario IV as Choi-Wan moves into the mid-latitudes more slowly and reaches a position related to the other scenarios 12 h later (20 September 2009, 12 UTC, Figure 5.35b). In scenario IV, Choi-Wan approaches the centre of a strong mid-latitude ridge over Japan rather than a trough that occurs in the other scenarios. At the onset of ET, the circulation pattern in scenario III is in accordance with the northwestern type, identified by Harr et al. (2000) as the trough west of Choi-Wan is the dominant flow feature. In two other scenarios the dominant flow structure is located east of Choi-Wan (the central to eastern mid-latitude trough in scenario II and the strong extratropical cyclone in scenario I), hence they resemble the northeastern circulation pattern, identified by Harr et al. (2000). Scenario IV also has a markedly trough in the eastern North Pacific, but the circulation pattern is dominated by the ridge north of Choi-Wan and the adjacent upstream trough over Russia.

The differences in the 500 hPa geopotential height and mean sea level pressure pattern are also manifested in the total flux (*totflux*) of K_e and its net *generation*. In scenario I, the comparatively strong extratropical cyclone southeast of Kamchatka clearly transports K_e towards the central Pacific trough (Figure 5.34c). Some total K_e flux is already directed from Choi-Wan towards the mid-latitude trough as well. Baroclinic conversion from eddy available potential into eddy kinetic energy ahead of the mid-latitude cyclone and its associated shortwave trough is comparatively strong (Figure 5.34e). At the same time this region of positive *bcli* is related to strongly diverging ageostrophic geopotential fluxes (Figure 5.34g), which directly disperse parts of this newly converted eddy kinetic energy towards the downstream K_e centre on the rear flank of the central Pacific trough. Hence, although strong baroclinic conversion suggest a clear increase of K_e,

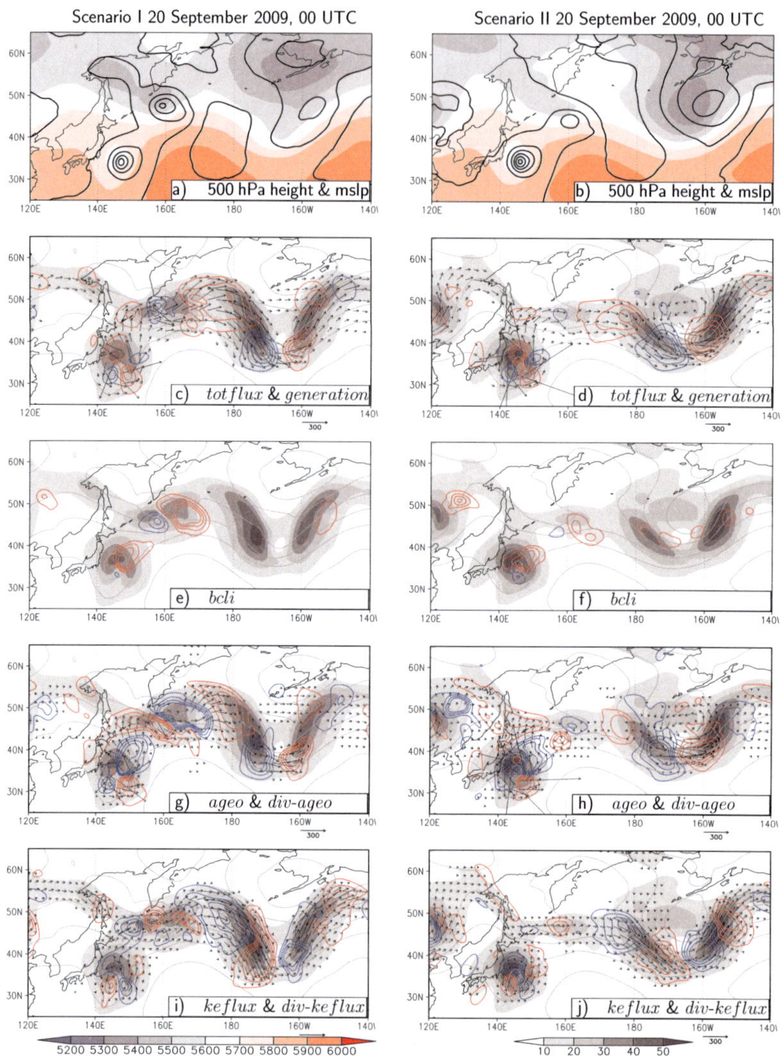

Figure 5.34: Scenario I (left) and II (right) at the onset of interaction between Choi-Wan and the mid-latitude flow. a, b) 500 hPa geopotential height (shaded) and mslp (contours); c, d) K_e (shaded), *totflux* (arrows) and *generation* (contours); e, f) K_e (shaded) and *bcli* (contours); g, h) K_e (shaded), *ageo* (arrows) and *div-ageo* (contours); i, j)K_e (shaded), *keflux* (arrows) and *div-keflux* (contours). Units, increments and colors as in Figure 5.27. See Table 3.1 for abbreviations.

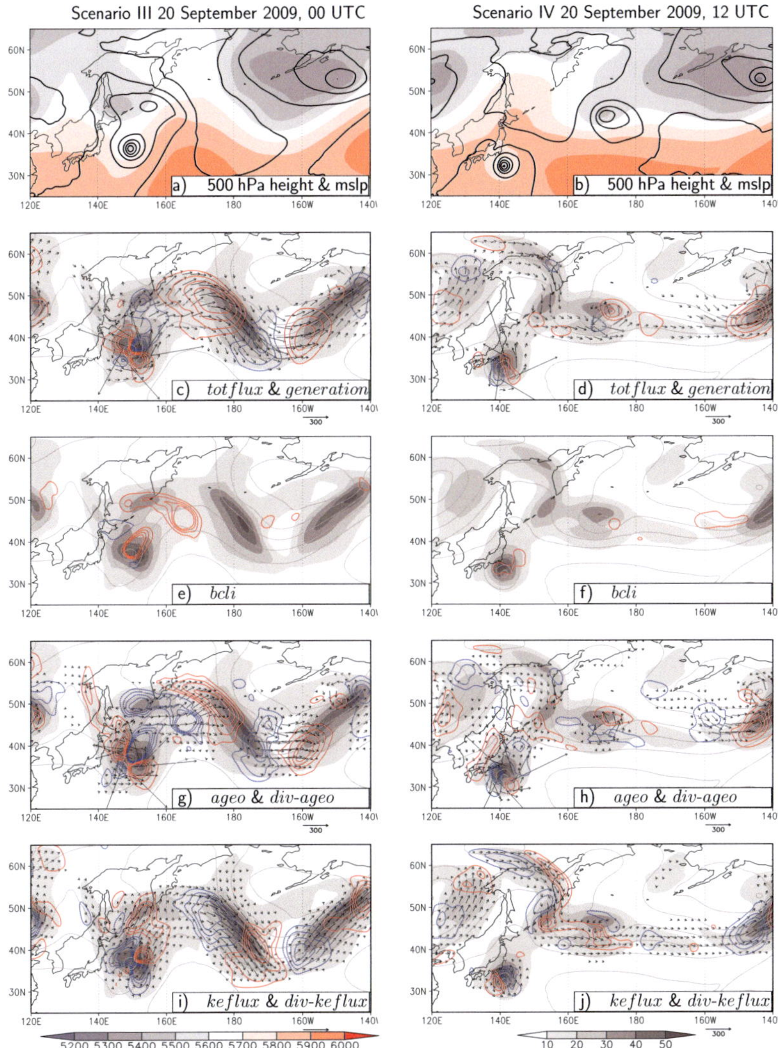

Figure 5.35: Scenario III (left) and IV (right) at the onset of interaction between Choi-Wan and the mid-latitude flow. a, b) 500 hPa geopotential height (shaded) and mslp (contours); c, d) K_e (shaded), *totflux* (arrows) and *generation* (contours); e, f) K_e (shaded) and *bcli* (contours); g, h) K_e (shaded), *ageo* (arrows) and *div-ageo* (contours); i, j) K_e (shaded), *keflux* (arrows) and *div-keflux* (contours). Units, increments and colors as in Figure 5.27. See Table 3.1 for abbreviations.

downstream dispersion of K_e via diverging ageostrophic geopotential fluxes (*div-ageo*) depletes the net *generation* in this region and the shortwave trough south of Kamchatka is not amplified appreciably. On the other hand, converging K_e fluxes and thus a gain in K_e coincide with an acceleration of the flow through the rear flank of the downstream trough and supports its further amplification (Figure 5.34i). Baroclinic conversion in the northeastern quadrant of Choi-Wan, due to rising warm air corresponds with diverging ageostrophic geopotential fluxes of equal strength, which diminish the net generation in this area. This contributes to a radiation of K_e towards the north into the K_e centre associated with the extratropical cyclone, where *div-ageo* converges between 140-160°E and 44-47°N (Figure 5.34g). However, negative baroclinic conversion on the rear side of this K_e centre, near 45°N, 158°E (Figure 5.34e) causes the net generation to be negative, other than might be expected from *div-ageo* (Figure 5.34c). Ageostrophic geopotential fluxes also recirculate K_e within the centre of Choi-Wan and cause the formation of a dipole structure of positive and negative generation (Figure 5.34c, between 30-40°N, 140-155°E). Flux of K_e by the total wind (*div-keflux*) is important for the downstream support as it predominantly redistributes eddy kinetic energy within the K_e centres. Hence, strong diverging *div-keflux* in the upstream edge and converging *div-keflux* on the downstream edge of the K_e maximum at the rear side of the central North Pacific trough redistributes K_e within this centre from the region of net generation to the region of net loss. The latter one is caused by already downstream dispersing ageostrophic geopotential fluxes towards the front side of this downstream trough. The zonal component of *div-keflux* further triggers the eastwards propagation of the both K_e centres and the associated central North Pacific trough. Overall at the onset of ET, scenario I is characterised by a clear downstream dispersion of K_e, originating in the pre-existing extratropical cyclone. Energy fluxes from Choi-Wan are directed towards this extratropical K_e centre and thus, albeit indirectly, support the observed downstream baroclinic development also. Accumulation of K_e in the downstream trough supports its further amplification.

Downstream fluxes, as well as the strength of the extratropical cyclone are weaker in scenario II (Figure 5.34d) at this time (20 September 2009, 00 UTC). Baroclinic conversion also generates K_e ahead of the weak mid-latitude cyclone, but to a lesser extent (Figure 5.34f). Downstream dispersion via *div-ageo* just slightly disperses this K_e towards the downstream trough (Figure 5.34h), leading to net *generation* of K_e at the upstream entrance region of this centre (Figure 5.34d). Furthermore, baroclinic conversion also occurs in the northeastern quadrant of Choi-Wan (Figure 5.34f). While parts of this eddy kinetic energy are recirculated around Choi-Wan by *div-ageo* and *div-keflux*, a large amount of K_e is also directly dispersed into the K_e centre extending northwards. A couplet of diverging and converging ageostrophic geopotential fluxes northwest of Choi-Wan between 50-55°N and 125-135°E indicates a slight upstream support of K_e in the northwestwards extending centre over Sakhalin (large island north of Japan). This may further amplify the trough north of Choi-Wan. Fluxes of K_e by the total wind (*keflux*) recirculate K_e within the centre of Choi-Wan. They also support further downstream baroclinic development, emanating from the broad central North Pacific trough, as K_e is redistributed from the rear to the front of the associated K_e maxima, on the western and eastern trough flanks. Weak downstream dispersion via *div-ageo* (Figure 5.34h) is already initiated in the Gulf of Alaska, as the flow in front of the central North Pacific trough is accelerated, which leads to a stronger ageostrophic wind in the adjacent ridge. Meanwhile, recirculation of K_e in the base of the strong central North Pacific trough, together with slight internal baroclinic conversion help the K_e centres of this trough to sustain. Eddy kinetic energy fluxes from Choi-Wan mainly supply the K_e centre to its north during the onset of ET in scenario II. Only minor energy transport exists between the centre north of Choi-Wan and the central North Pacific wave pattern. Hence, although the onset of the transitioning storm is closely related in scenario I and II during the early stages of ET, the weaker mid-latitude disturbance in scenario II and the reduced generation and dispersion of eddy kinetic energy diminished the influence of the transitioning storm on the downstream baroclinic development.

In scenario III strong baroclinic conversion occurs along a baroclinic zone which extends from the extratropical cyclone to the southeast. The northeastward directed flow in the eastern flank of the transitioning cyclone and the associated upper-level trough induces ascent of warm air in the eastern part of the baroclinic zone, causing a maximum in baroclinic conversion near 162-170°E, 42-49° (Figure 5.35e). This contributes to a further amplification of the downstream ridge. Strongly divergent ageostrophic geopotential fluxes occur in conjunction with the baroclinic conversion and support downstream baroclinic development (Figure 5.35c, e, g, i). This overlap of *bcli* and *div-ageo* leads to a region of net loss of K_e over the Kuril Islands (between 45-52°N and 150-157°E, Figure 5.35c). In contrast, strong net *generation* (east of Kamchatka) is associated with the eastern part of the baroclinic zone and the downstream centre in the crest and on the front flank of the ridge just east of Choi-Wan. This generation of K_e coincides with an acceleration of the flow through the ridge and supports the amplification of the wave pattern. Hence, this K_e centre also acts as a source for further downstream development towards the West Coast of North America, due to *div-keflux* (advection and redistribution) and *div-ageo* (dispersion) in the downstream trough. The K_e centre of Choi-Wan starts to merge with the K_e centre of the extratropical cyclone, recognisable due to enhanced K_e between the two maxima (45-50°N, 145-155°E, Figure 5.35c) and a dipole flux pattern emanating from Choi-Wan towards the mid-latitude centre (35-55°N, 140-160°E, Figure 5.35i). Baroclinic conversion is also strong in the northeastern quadrant of Choi-Wan (Figure 5.35c) but related to divergent ageostrophic geopotential fluxes which disperse parts of the newly converted K_e from Choi-Wan into the northern and western part of the centre, associated with the extratropical cyclone and also towards the downstream centre. Thus, Choi-Wan also contributes clearly to downstream baroclinic development. Concomitantly, *div-ageo* also acts to recirculate K_e within the centre of Choi-Wan. The significantly strong baroclinic conversion at the baroclinic zone and in the transitioning Choi-Wan, together with the redistribution of K_e between the centres of Choi-Wan and the extratropical disturbance, apparently support the extratropical centre against its loss of K_e, due to downstream radiation towards the central North Pacific trough. Downstream transport of K_e, emanating from Choi-Wan and the extratropical cyclone clearly strengthens the amplification of the downstream wave pattern. An only minor support from the upstream K_e centre affects the region around Choi-Wan at this time.

The onset of interaction between Choi-Wan and the mid-latitude flow is different in scenario IV, as the first clear interaction occurs 12 h later (20 September 2009, 12 UTC), compared to the cases discussed previously. In addition to this time shift the position of Choi-Wan relative to the mid-latitude flow distinguishes this scenario from the others. Choi-Wan is located close to Japan, almost on the rear flank of the strongly amplified ridge over the western North Pacific (Figure 5.35b). Total fluxes of K_e are rather weak in association with Choi-Wan, but are stronger through the crest of the amplified ridge north of Choi-Wan (Figure 5.35d). The ridge hinders strong baroclinic conversion of K_e in Choi-Wan, but generation of K_e due to baroclinic processes is also weak in the extratropical cyclone (Figure 5.35f). However, parts of eddy kinetic energy from Choi-Wan are then dispersed via *div-ageo* through the crest of the ridge towards the K_e centre on the rear flank of the trough, associated with the extratropical cyclone in the central North Pacific (Figure 5.35h).

As most of the K_e in this downstream cyclone is recirculated, and the flow is rather zonal, inhibiting strong ageostrophic fluxes, no clear downstream baroclinic development is observed to emanate from the western towards the eastern North Pacific. Advection of K_e by the total wind recirculates K_e within the centre of Choi-Wan and redistributes K_e in the crest of the ridge. In this case, the albeit weak baroclinic conversion in Choi-Wan forms an additional source of K_e for the mature downstream cyclone. However, due to a weakly amplified wave pattern and thus only weak ageostrophic flow, a downstream baroclinic development cannot be established over the central North Pacific at this time.

During the subsequent 24 hours, the development starts to differ more strongly between the four scenarios. In scenario I, the extratropical cyclone couples with the dominant, nearly stationary trough over the Gulf of Alaska (Figure 5.36a), and broadens until 21 September 2009, 00 UTC. Its associated K_e centre is elongated between 160-180°E, 45°N and thus to the east-northeast of Choi-Wan. Meanwhile, the transport of K_e towards the western flank of the downstream trough weakens (Figure 5.36c), reducing the potential for further amplification of the downstream trough. Strong ageostrophic geopotential fluxes, diverging from the northeastern side of Choi-Wan, where some positive baroclinic conversion provides an additional source of K_e (Figure 5.36e), disperse eddy kinetic energy into the western parts of the K_e centre of the extratropical cyclone (Figure 5.36g). But negative baroclinic conversion to the rear of this extratropical K_e centre, due to sinking warm air masses in the left entrance region of an upper-level wind maxima (not shown) diminishes K_e generation there. Simultaneously, ageostrophic geopotential fluxes diverge from the upstream K_e centre along the coast of Russia and converge on the western flank of Choi-Wan's K_e centre (140°E, 40°N) causing net *generation* there. They also indicate a pattern of downstream dispersion in the ridge north of Choi-Wan. Fluxes of K_e by the total wind (*keflux*) mainly act to transport eddy kinetic energy from the rear to the front of the extratropical K_e centre (Figure 5.36i). In the eastern North Pacific, downstream baroclinic development apparently occurs near the West Coast of North America. Strong dispersion of K_e from the rear to the front flank of the downstream trough (Figure 5.36g) accelerates the flow along the front flank of the trough, supporting further amplification of the ridge over the West Coast. During the ET in scenario I the strong downstream support emanating from Choi-Wan and the extratropical cyclone is decreased, compared to the earlier forecast day. Baroclinic conversion of K_e is reduced in both cyclones. Ageostrophic geopotential fluxes still diverge from the K_e centre of Choi-Wan and support the centre associated with the extratropical cyclone. The remnants of Choi-Wan gain additional K_e from convergent ageostrophic geopotential fluxes that emanate from the upstream centre over Russia.

In scenario II, a distribution of the 500 hPa geopotential height and surface pressure field related to scenario I is reached 12 hours later (21 September 2009, 12 UTC (Figure 5.36b). However, the strength of the two cyclonic systems is weaker, as are the magnitudes of the terms in the K_e budget. The total flux of K_e just indicates a slight downstream spread of K_e from Choi-Wan and the extratropical cyclone towards the dominant trough in the eastern North Pacific. On the other hand, K_e centres associated with this trough, especially at its front flank act as a mature source for downstream baroclinic development over western North America (Figure 5.36d, Figure 5.36h). Weak baroclinic conversion on the eastern side of the extratropical cyclone (Figure 5.36f) is superimposed by diverging ageostrophic geopotential fluxes (Figure 5.36h). Hence, only minor net *generation* is associated with this cyclonic system (Figure 5.36d). A divergence-convergence couplet of *div-ageo* between the remnants of Choi-Wan and the upstream K_e centre, near 140-160°E and 40-50 °N indicate this upstream centre to maintain the K_e in Choi-Wan, as baroclinic conversion causes loss of K_e in this region. No energy is dispersed from Choi-Wan towards the extratropical cyclone. In addition to its well-known role in downstream development in the eastern part of the Northern Pacific, K_e flux with the total wind (*keflux*) transports eddy kinetic energy through the ridge north of Choi-Wan and, albeit weakly towards the dominant downstream trough (Figure 5.36j). The weaker K_e fluxes in scenario II coincide with an overall decreased magnitude in K_e maxima, especially in the western North Pacific. Although the centre of Choi-Wan gains support from upstream regions, no strong downstream support emanates from Choi-Wan into the mid-latitudes. The extratropical cyclone is also less conducive to downstream development.

As Choi-Wan merges completely with the extratropical cyclone between 00 UTC 20 September and 00 UTC 21 September in scenarion III, the development on 12 UTC 20 September is shown in addition, to provide further insight into the merger. Choi-Wan propagates towards the extratropical cyclone and is located ahead of the cut-off low over the Kuril Islands between Japan and Kamchatka (Figure 5.37). The K_e centres of the two cyclonic systems start to merge, while *totflux* transports K_e through the adjacent ridge towards the downstream trough in the central North Pacific. Baroclinic conversion of K_e is still strong along the baroclinic zone south and southeast of Kamchatka and in the northern parts of the transitioning typhoon (Figure 5.37b). However, ageostrophic geopotential fluxes disperse K_e out of the region of baroclinic conversion near Choi-Wan towards the upstream end of the K_e centre on the western flank of the downstream trough (Figure 5.37d), leading to net loss of K_e in the northern quadrant of Choi-Wan (Figure 5.37b, 157-162°E, 42-47°N). On the other hand, divergent ageostrophic geopotential fluxes in the K_e centre of the extratropical cyclone (Figure 5.37d, 150-157°E, 45-55°N) recirculate K_e in the merging centre towards the southwest, where they accumulate and cause net *generation* (Figure 5.37b, 148-161°E, 35-45°N). The flux of K_e with the total wind mainly redistribute eddy kinetic energy from regions of generation

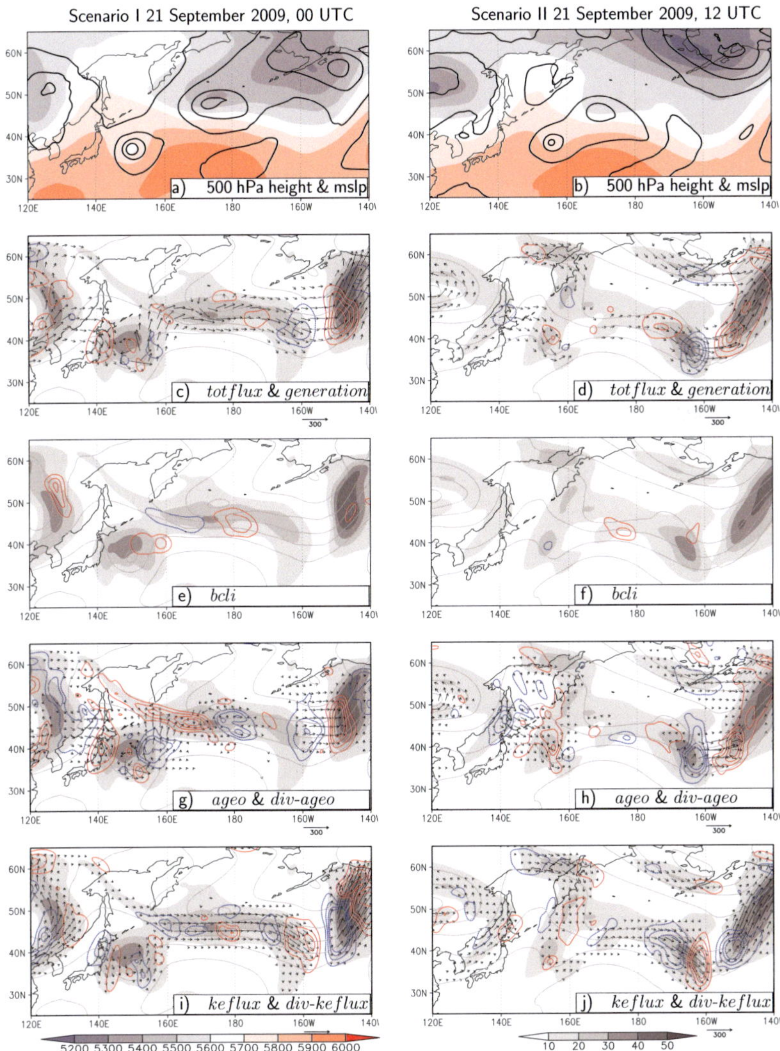

Figure 5.36: As Figure 5.34, but around the onset reintensification.

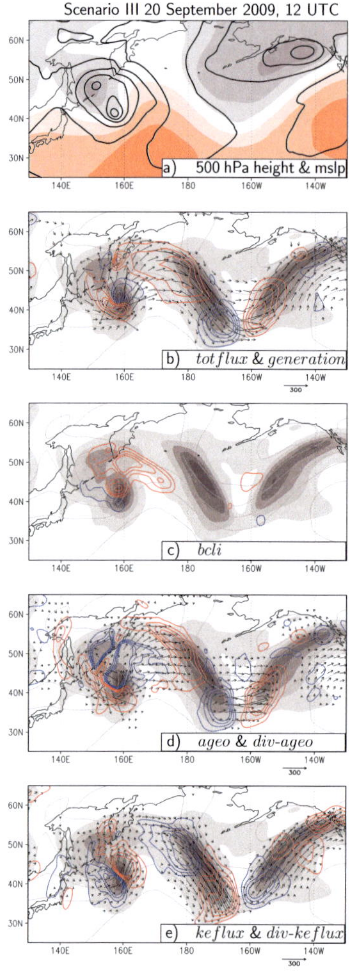

Figure 5.37: Scenario III during the merger between Choi-Wan and the extratropical cyclone. Plots as in Figure 5.35.

towards regions with loss of K_e in the centre of Choi-Wan and in the downstream trough.

During the subsequent 12 hours, the Choi-Wan and the extratropical cyclone merged completely to form a strongly reintensified system on 21 September 2009, 00 UTC (Figure 5.38a). Baroclinic conversion in the baroclinic zone and advection of K_e during the past 12 h accumulated K_e in the northern and eastern portion of the reintensified cyclone. A clear downstream development situation, enclosing the remnants of Choi-Wan, is indicated by the total flux of K_e (Figure 5.38c). Strong baroclinic conversion of available potential into eddy kinetic energy is stretched out over the northern part of the K_e centre, associated with the reintensified Ex-Choi-Wan (Figure 5.38e). Thereby, the rate of baroclinic conversion still clearly exceeds the contribution of $bcli$ in the other scenarios. This area is partially superimposed by strongly diverging ageostrophic geopotential fluxes that disperse K_e over and through the ridge towards the downstream trough in the central North Pacific. The accumulation of K_e in the trough flanks coincides with

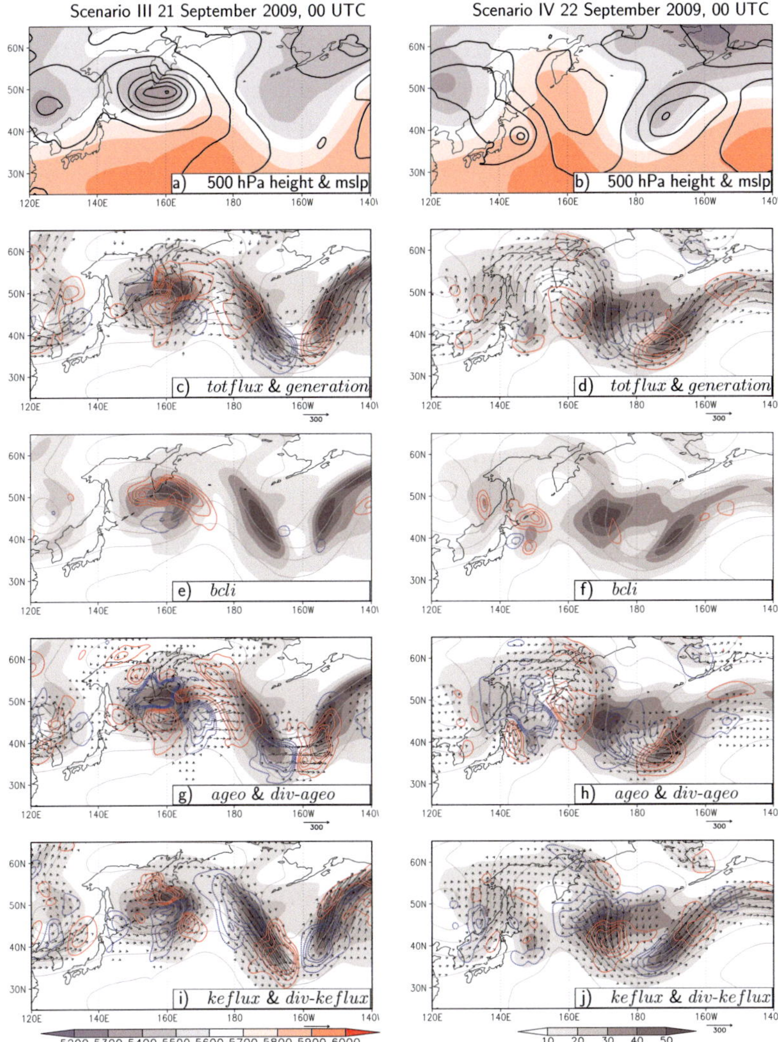

Figure 5.38: As Figure 5.35, but around the completion (III) and the onset (IV) of reintensification.

an acceleration of the flow through the downstream trough and hence supports its further amplification. Simultaneously, the subsequent downstream baroclinic development is enhanced also. Recirculating fluxes of K_e and the associated divergence and convergence pattern in *div-ageo* and *div-keflux* act to sustain the K_e centre of Ex-Choi-Wan as they retain parts of the baroclinically converted eddy kinetic energy in the K_e centre of the transitioning storm. The interplay between strong baroclinic conversion and recirculating K_e fluxes during the ET act in conjunction with the downstream dispersion of K_e to maintain the K_e centre of Choi-Wan and contribute to downstream development.

The different development of scenario IV is enhanced further on 22 September 2009, 00 UTC. Strong total flux of K_e over and through the ridge north of Choi-Wan enhance K_e on the rear side of the downstream trough, contributing to its amplification. Strong baroclinic generation in the northern part of the remnants of Choi-Wan (Figure 5.38f) is superimposed by diverging ageostrophic geopotential fluxes, which disperse K_e away from Choi-Wan through the ridge towards the upstream centre of the downstream trough (near 158-162°E, 40-50°N, Figure 5.38h). Thereby, the ageostrophic geopotential flux establishes a direct connection between the remnants of Choi-Wan and the adjacent downstream centre, crossing through the mature ridge, while only minor parts of the ageostrophic geopotential flux propagate over the top of the ridge towards the downstream centre. Subsequent downstream dispersion via *div-ageo* through the downstream trough into the K_e centre at its front already preconditions further downstream baroclinic development towards North America. In this case, moderate baroclinic conversion north of the rather weak remnants of Choi-Wan act as an additional energy source for the growing K_e centres, bordering the downstream trough, and support further amplification of this feature.

During the next 24-48 hours, the intensity of the remnants of Choi-Wan decreased while the extratropical cyclone, if it was not merged with Choi-Wan as in scenario III, is amplified further during its propagation into the eastern North Pacific in two of the three cases. On 23 September 2009, 00 UTC, only weak remnants of Choi-Wan can be identified in scenario I at a western position (153°E, 40 °N), while the extratropical ridge in the north has passed Ex-Choi-Wan. Hence, the transitioned storm is now located on the upstream flank of this ridge (Figure 5.39a). The shifting of the ridge north of Choi-Wan results from downstream dispersion of K_e via ageostrophic geopotential fluxes. This process had started on 21 September 2009, 00 UTC (130-140°E, 50-60°N, Figure 5.39g) and intensified during the subsequent forecast hour (not shown). Now, a broad, but rather weak trough is established over the central and eastern North Pacific. The extratropical cyclone has strengthen moderately and is located in the Gulf of Alaska (Figure 5.39a). The total fluxes of K_e suggest a downstream radiation of K_e, emanating from the ridge northeast of Choi-Wan (Figure 5.39c). Some baroclinic conversion is associated with a slight K_e centre, extending from Choi-Wan towards the northwest (155°E, 45°N, Figure 5.39e). However, diverging ageostrophic geopotential fluxes disperse most of this K_e directly towards the southwest of Choi-Wan's energy maximum (cf. convergence region near 40°N, 145°E, Figure 5.39g). To a weaker extent, K_e is also dispersed from the northern parts of the centre of Choi-Wan into the northern K_e maxima east of Kamchatka (Figure 5.39g). Flux of K_e by the total wind transports K_e from the *div-ageo*-convergence area in the upstream entrance region of the centre on the rear flank of the downstream trough towards the bottom of this trough and the exit region of the associated K_e centre (Figure 5.39i). From there, ageostrophic geopotential fluxes support further downstream propagation of K_e. Weak baroclinic conversion is associated with the extratropical cyclone in the Gulf of Alaska (Figure 5.39e), but recirculating K_e fluxes help to sustain this centre. However, ridge breaking over the North American West Coast led to formation of a cut-off cyclone and a broad but weakly amplified ridge over the continent.

In scenario II, the remnants of Choi-Wan are located ahead of a slight short-wave trough with a tilted axis (from 164°E, 36°N to 170°E,42°N) on 23 September 2009, 12 UTC, and thus on the opposite side of the ridge to scenario I. The extratropical cyclone has intensified and is embedded in a dominant trough in the central to eastern North Pacific (160 °W Figure 5.39b). This intensification was supported by strong energy fluxes through the crest of the ridge in the western North Pacific 12 h earlier, while no energy fluxes emanated from Choi-Wan (not shown). The downstream dispersion of K_e in scenario II drops completely over the western to central North Pacific (170°E - 170°W), while downstream dispersion towards North America is reduced (Figure 5.39d), compared to the former forecast time. Baroclinic conversion is rather weak in both cyclonic systems (Figure 5.39f). Ageostrophic geopotential fluxes (Figure 5.39h) are now enhanced over the western North Pacific, they emanate from upstream regions and accumulate K_e in centres around the shortwave trough near Choi-Wan. This enables further amplification of the trough and thus allows a reintensification of Choi-Wan towards later forecast times (see Discussion). In the K_e centre of the extratropical cyclone, ageostrophic geopotential fluxes act to recirculate K_e, without any downstream dispersion. Fluxes of K_e by the total wind, on the other hand transport K_e from this cyclone through the ridge over the West Coast of North America, towards a shortwave disturbance west of Vancouver Island and even further downstream (Figure 5.39j). At this time, the downstream system seems to be completely unconnected from the remnants of Choi-Wan in scenario II. However, Ex-Choi-Wan is located ahead of a shortwave trough disturbance, whose growth is supported by energy fluxes from upstream regions.

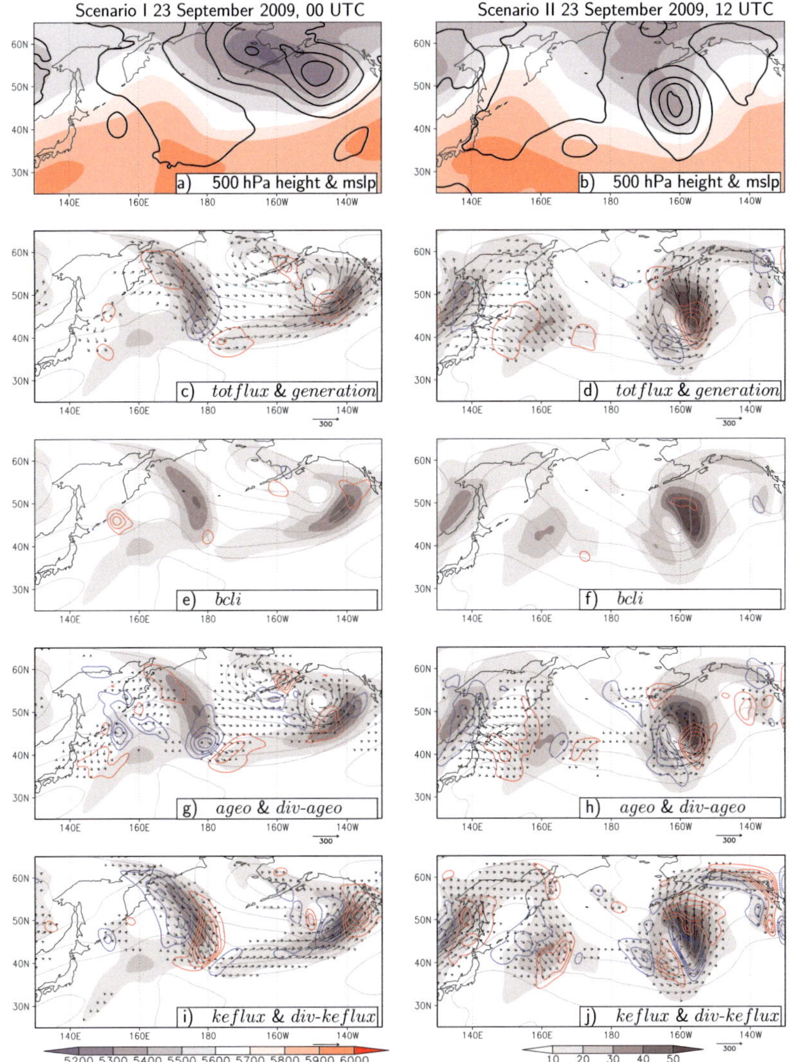

Figure 5.39: As Figure 5.35, but after completion of ET.

The strongly reintensified system that formed from Ex-Choi-Wan and the extratropical cyclone in scenario III connects to the dominant trough in the eastern North Pacific, while surface pressure of the merger weakened during the next 24 hours (23 September 2009, 00 UTC, Figure 5.40a). Total energy fluxes highlight only minor downstream dispersion of K_e from Choi-Wan towards the centre in front flank of the the dominant downstream trough. Meanwhile, energy fluxes towards North America exist in the eastern North Pacific (Figure 5.40c). A couplet of ageostrophic geopotential flux divergence and convergence, emanating from the K_e centre over Japan, indicate upstream support into the K_e centre at the southern flank of the merger of Choi-Wan and the extratropical cyclone. This region is superimposed by weak baroclinic destruction (40 W/m^2), which is below contour level (50 W/m^2), but acts a sink for K_e by descending warm air masses in upper levels (not shown). As ageostrophic fluxes near Choi-Wan mainly act to recirculate K_e, no evident downstream support towards the dominant eastern North Pacific trough exists. This lack of upstream support causes the K_e centre at the rear flank of the downstream trough to be further weakened. In addition, ageostrophic geopotential flux divergence disperses baroclinically converged K_e from the rear to the front flank in the bottom of the downstream trough (Figure 5.40e,g). Baroclinic conversion near the exit region of the K_e centre ahead of the downstream trough acts as an, albeit weak, source of K_e for the downstream K_e centre over central North America to where it is dispersed via *div-ageo* (near 140°W, 55°N, Figure 5.40g). Flux of K_e with the total wind (*keflux*) near Ex-Choi-Wan causes redistribution in this centre and also contributes to its eastwards propagation through advection (Figure 5.40i). In this scenario, Ex-Choi-Wan has now an only minor influence on downstream dispersion of K_e. In the K_e centres, surrounding the downstream trough a clear downstream baroclinic development pattern exists, while the K_e centre on the rear flank is already prone to decay, as no support from Choi-Wan and no internal baroclinic conversion help it to sustain, while it radiates K_e to the front flank.

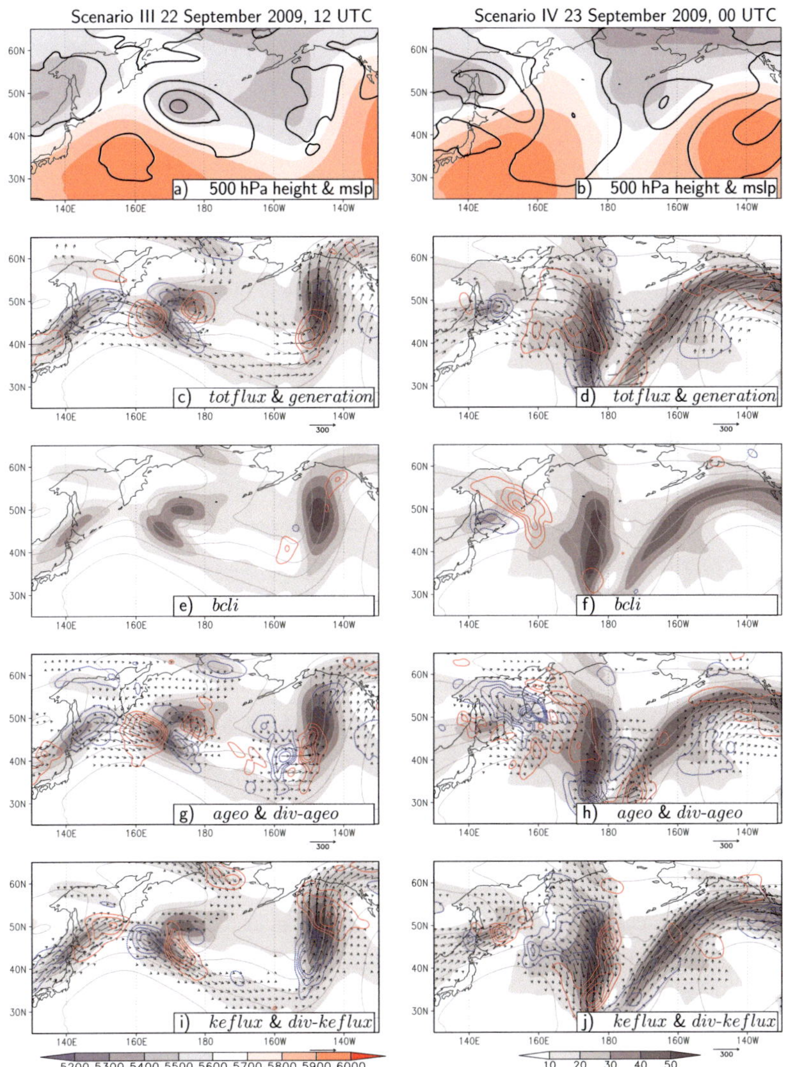

Figure 5.40: As Figure 5.35, but after completion of ET.

In scenario IV the weak remnants of Choi-Wan have moved northwards and been absorbed in a broad but weak low pressure centre, existing over Sakhalin (136-145°E, 50-55°N, Figure 5.40b). The downstream wave pattern is strongly amplified and the total flux of K_e clearly resembles a pattern of downstream baroclinic development (Figure 5.40b,d). The extratropical cyclone ahead of the downstream trough has not intensified further, though. Comparatively strong baroclinic conversion of K_e southwest of Kamchatka is associated with the baroclinic zone, elongating northeast of the trough and the remnants of Choi-Wan (Figure 5.40f). However, the superimposed ageostrophic geopotential flux divergence disperses K_e from this region towards the adjacent centre in the rear flank of the mature downstream trough, causing net generation there (160-170°W, 40-60°N, Figure 5.40h). But in turn, the centre near the remnants of Choi-Wan experiences net loss of K_e near 56°N and 147°E (Figure 5.40d). The net generation in the K_e centre at the rear trough flank accelerates the flow and hence supports further amplification of this mature trough in the central North Pacific. In the base of this trough, ageostrophic geopotential fluxes mainly act to recirculate K_e. Hence, the downstream dispersion from the base of the trough is weakened, while K_e from the centre between the mature trough and a subtropical high pressure system to its southeast (near 155°W, 36-46 °N) is dispersed via *div-ageo* over the ridge towards the West Coast of North America. Fluxes of eddy kinetic energy with the total wind mainly indicate the well-known advection pattern inside the K_e centres, causing their downstream propagation. In this scenario, the baroclinic conversion near the remnants of Choi-Wan apparently acts as an additional source of eddy kinetic energy for the K_e centre of the downstream trough, enabling its further amplification. In turn, the K_e centre near Choi-Wan will decay further, because of the net loss of K_e. Due to strong recirculation of ageostrophic geopotential fluxes the K_e centres, bordering the mature trough will be partially sustained. The recirculation cause the formation of a cut-off low during the subsequent forecast hours.

The analysis of the ensemble forecast shows that the ET of Typhoon Choi-Wan led to four completely different development scenarios. The final outcome of these scenarios and their impact on the downstream region, in this case North America, will be introduced in the summary.

c) Summary and Discussion for Choi-Wan

The analysis of the EKE-budget for a number of distinct development scenarios for Typhoon Choi-Wan reveals the role of the transitioning cyclone in the various synoptic developments. During the early stages of the ET of Choi-Wan, the mid-latitude circulation patterns already differ between the four forecast scenarios. This also affects the ET of Choi-Wan and its influence on the mid-latitude flow. In one case (scenario I), Choi-Wan moves towards a moderate upper-level trough, while a comparatively strong extratropical cyclone northeast of Choi-Wan dominates the mid-latitude flow (northeast circulation pattern, described by Harr et al., 2000). Strong baroclinic conversion from eddy available potential into eddy kinetic energy occurs in the extratropical cyclone and in the transitioning Choi-Wan, but the regions of baroclinic conversion are separated from each other. Energy fluxes clearly act to transport K_e from the centre of Choi-Wan towards the kinetic energy centre of the extratropical cyclone as the outflow of Choi-Wan crosses isohypses towards lower heights. Fluxes from the extratropical cyclone through the crest of the downstream ridge disperse K_e towards the dominant trough in the eastern North Pacific. Hence, in this case the two cyclones support downstream baroclinic development in the framework of Orlanski and Sheldon (1995) and contribute to the further amplification of the downstream trough.

Another forecast scenario (II) shows a related set-up in the circulation pattern during the onset of ET, but the dominant flow feature in this case is the downstream trough in the eastern North Pacific, while the extratropical cyclone northeast of Choi-Wan is weaker. Baroclinic conversion in Choi-Wan and eddy kinetic energy fluxes from Choi-Wan towards the K_e centre of the extratropical cyclone to its northeast are seen as in scenario I. However, less baroclinic conversion is associated with the weak extratropical cyclone in scenario II, compared to scenario I. The same applies to the transport of K_e towards the downstream trough in the eastern North Pacific, which is clearly reduced in scenario II along the weaker geopotential gradient west and north of the downstream ridge. Hence, Choi-Wan and the extratropical cyclone are less conducive to downstream baroclinic development in this case.

The third forecast scenario (scenario III) depicts the movement of Choi-Wan into a rather northwestern circulation pattern, as the transitioning storm moves ahead of an upper-level trough. Significantly strong baroclinic conversion occurs in the northern quadrant of Choi-Wan and in a baroclinic zone extending southeastwards from the extratropical cyclone. Thereby, maximum baroclinic conversion in the southeastern part of the baroclinic zone is caused by ascending warm air masses, driven by the northeastward circulation of the transitioning cyclone. Energy fluxes transport K_e from Choi-Wan into the extratropical cyclone as the upper-level outflow of Choi-Wan crosses towards lower heights. The extratropical cyclone, in turn, acts as a source for intensification of the downstream baroclinic development in the central and eastern north Pacific. Baroclinic conversion and energy fluxes in this case somewhat resemble the pattern found in scenario I but are more intense. In contrast to the rather separated cyclones in scenario I, the northeastward flow around the transitioning cyclone clearly enhances baroclinic conversion in scenario III. Overall, a strong support for downstream baroclinic development emanates from Choi-Wan.

A completely different circulation pattern during the onset of ET is found in the fourth scenario (scenario IV). Choi-Wan becomes embedded in the dominant ridge over Japan, and hence in a rather unfavourable position for reintensification. Baroclinic conversion is weak due to the reduced baroclinicity within the ridge. Nevertheless, moderate energy fluxes emanate from Choi-Wan through the ridge towards the K_e centre near the extratropical cyclone, while a clear transport of K_e through the crest of the ridge over Japan into this centre exists also.

The next ET step under consideration is reached about 24-36 h later (around clustering time) in the distinct scenarios and the differences in the synoptic development are further enhanced. The different forecast times at which this ET stage is reached in the several scenarios are in accordance with the identification of a time line in the several scenarios during the clustering process. At this time, the strong extratropical cyclone in scenario I connects to the downstream trough and becomes more separated from the remnants of Choi-Wan. Although Choi-Wan still disperses K_e towards the extratropical cyclone, energy fluxes have decreased substantially compared to the earlier day. Choi-Wan is now more located in the upstream ridge, this reduces cross contour flow towards lower pressure and thus the dispersion of K_e. Baroclinic conversion is also reduced. Hence, Choi-Wan's impact on downstream baroclinic development is now reduced. The energy centre near Choi-Wan experiences slight maintenance from upstream regions due to cross-contour flow through the upstream ridge. This scenario represents the cluster that was still rather tropical during the clustering, fitting to result that the occurrence of Choi-Wan in the provided fields did not change strongly between the actual and the earlier forecast times, at least in comparison with the changes in the other scenarios.

In contrast, the transitioning Choi-Wan weakened considerably in scenario II (the development in scenario II was depicted 12 h later than in scenario I, but weakening of Choi-Wan started earlier). Energy fluxes from Choi-Wan towards the weak extratropical cyclone and also further downstream are depleted. On the other hand, the K_e centre near Choi-Wan gains additional K_e from the upstream centre over Russia by cross-contour flow. Scenario

II was included in the decaying cluster, which fits quite well with the described synoptic development of scenario II around the clustering time.

Choi-Wan merges with the extratropical cyclone and the whole system reintensifies strongly in the third scenario (scenario III). Strong baroclinic conversion due to ascending warm air in the northern quadrant of Choi-Wan is maintained. Thus, the mature cyclone has still a significant contribution to downstream baroclinic development in the eastern North Pacific by diverging ageostrophic geopotential fluxes. Consistent with this development, scenario III is a representative member for the reintensifying cluster.

In the fourth scenario, Choi-Wan is now located ahead of the dominant upstream trough, but the strong ridge over Japan still inhibits a potential reintensification of Choi-Wan. However, eddy kinetic energy from Choi-Wan and the upstream centre is transported through and over the crest of the dominant ridge and support the further amplification of the downstream trough in the eastern North Pacific. Baroclinic conversion occurs north of Choi-Wan, where the circulation of the transitioning storm causes the ascent of warm air masses along the baroclinic zone north of Choi-Wan. Following the idea of the representative members, scenario IV should capture a development in which the ET of Choi-Wan is still under way. At this time (which is 24 h after the clustering time), Choi-Wan interacts with the mid-latitude flow but appears to be still during its ET, maybe postponed by the strongly amplified ridge.

During the next forecast days, the remnants of Choi-Wan first weakened in all scenarios, but are in distinct phasing to the mid-latitude wave pattern, which support a slight reintensification in some of the scenarios. Towards the end of the forecast the synoptic situation (e.g. precipitation and surface temperatures) downstream of the ET event differs clearly between the four scenarios. This emphasises the possible impact of an ET event for the synoptic development in downstream regions and on the other hand points to loss of predictability during the interaction of a tropical cyclone with the mid-latitude flow. In the first scenario, the ridge passed Choi-Wan to its north. Energy fluxes emanate from the front flank of this ridge towards the broad trough in the eastern North Pacific, which contains the further amplified extratropical cyclone. Choi-Wan is located on the upstream side of the ridge and its downstream dispersion of K_e of Choi-Wan has stopped. Towards the end of the forecast, the ridge downstream of Choi-Wan is eroded and the transitioned typhoon couples with a shortwave trough (Figure 5.41a, 150°E, 47°N). No clear energy centre or generation of K_e are associated with the remnants of Choi-Wan in the overall rather zonal mid-latitude flow (Figure 5.41c). Baroclinic conversion and divergent ageostrophic geopotential fluxes during the onset of reintensification twelve hours earlier eroded the K_e centre of Choi-Wan completely. Due to the lack in K_e fluxes from upstream regions, downstream development in the eastern North Pacific towards North America will be decreased further, but this is beyond the forecast range. The extratropical cyclone decays in the eastern North Pacific. A broad ridge in the temperature and geopotential height fields extends over western and central North America, accompanied with comparatively high surface temperatures over the western and central parts of North America (Figure 5.41e) and only minor precipitation over the US (Figure 5.41h). Precipitation maxima are associated with the decaying extratropical cyclone and the reintensifying Choi-Wan in the western North Pacific (Figure 5.41h).

In scenario II, Choi-Wan is still positioned east of the ridge over Japan and lies now ahead of a shortwave trough. Upstream energy fluxes due to cross-contour flow towards lower heights into this trough support the further reintensification of Choi-Wan until the end of the forecast time (Figure 5.41b,d). The former weak extratropical cyclone intensifies strongly as it moves towards the Gulf of Alaska ahead of another shortwave trough.

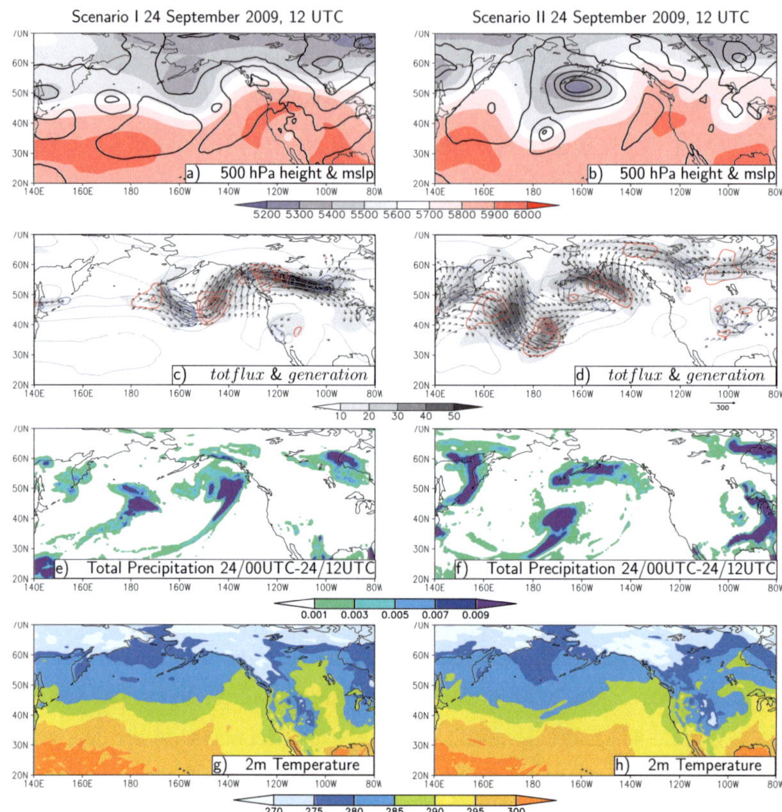

Figure 5.41: Distinct synoptic fields towards the end of the investigated forecast period for development scenarios for Typhoon Choi-Wan (valid at 24 September 2009, 12 UTC. a,b) 500 hPa geopotential height (shaded, in gpm) and mslp (contours, in 5 hPa increments) for scenario I and II, c,d) total eddy kinetic energy, *generation* and *totflux*, as Figure 5.34 for scenario I and II e,f) sum of large scale and convective precipitation (in m) between 24 September 2009, 00 UTC and 12 UTC scenario I and II, g,h) 2 m Temperature (in K) for scenario I and II

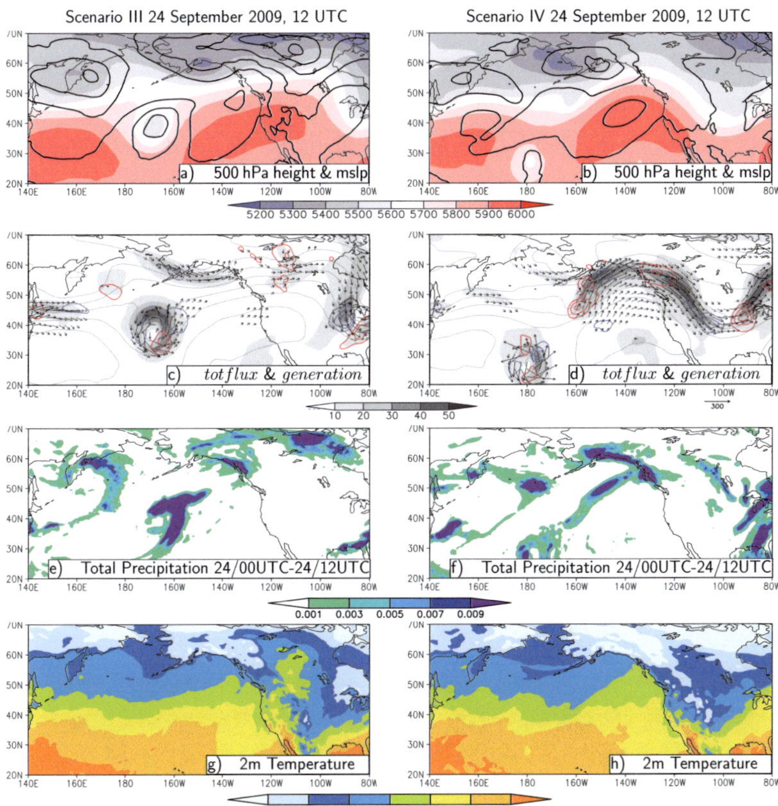

Figure 5.42: Distinct synoptic fields towards the end of the investigated forecast period for development scenarios for Typhoon Choi-Wan (valid at 24 September 2009, 12 UTC. a,b) 500 hPa geopotential height (shaded, in gpm) and mslp (contours, in 5 hPa increments) for scenario III and IV, c,d) total eddy kinetic energy, *generation* and *totflux*, as Figure 5.34 for scenario III and IV, e,f) sum of large scale and convective precipitation (in m) between 24 September 2009, 00 UTC and 12 UTC scenario III and IV, g,h) 2 m Temperature (in K) scenario III and IV

Wave breaking over central North America led to the formation of a cut-off cyclone over the Great Plains (Figure 5.41b), with low surface temperatures over and east of the Rocky Mountains (Figure 5.41g) and strong precipitation in the eastern parts of the US (Figure 5.41f). Hence, surface near temperatures in the Great Plains and the Great Lakes differ about 10 K between scenario I and II, and no precipitation over the US in scenario I opposes a band of comparatively strong precipitation over the eastern US in scenario II.

The strongly reintensified cyclone in scenario III weakens during the later forecast times. Its energy support towards downstream regions depletes as energy fluxes start to recirculate, while the cyclone still gains additional K_e due to cross-contour flow from upstream regions. The mature system separates from the mid-latitude wave pattern and finally forms a cut-off low in the central North Pacific until the end of the forecast (Figure 5.42a). Its former downstream support led to the formation of a strongly amplified ridge over western North America, which already weakens until the end of the forecast (Figure 5.42c). A sharp subsequent downstream trough is located over the Great Lakes. While temperatures in the ridge over western parts of North America are comparatively high, the sharp trough coincides with low temperatures east of the Rocky Mountains (Figure 5.42g), up to 15 K less than in scenario I and II. Precipitation occurs ahead of the southern portion of the trough over the southeastern US (Figure 5.42e).

In the fourth scenario (scenario IV), the remnants of Choi-Wan merged with another extratropical cyclone to its northwest over Sakhalin. Strong baroclinic conversion ahead of this merger acts as an additional eddy kinetic energy source for the further amplification of the downstream wave pattern, as K_e fluxes transport energy into the central North Pacific trough. Towards the end of the forecast the merged cyclone decays (Figure 5.42b), as does its downstream support (Figure 5.42d) and the baroclinic conversion (not shown). The trough in the central North Pacific cuts off and the remnants of the strong Rossby wave train propagate over North America, establishing a broad trough over central North America. This trough initiates precipitation over the eastern part of North America (Figure 5.42f) and rather low surface temperatures in the western part (Figure 5.42h), which opposes the temperature distribution in the western US, found in scenario I-III.

The comparison of the four forecast scenarios for the ET of Choi-Wan highlighted the various differences that can occur during the interaction between the transitioning tropical cyclone and the mid-latitude flow pattern, which pose a serious problem in predicting ET events in numerical models. The transitioning tropical cyclone may act as an additional source of eddy kinetic energy and contribute to downstream baroclinic development. Thereby the dipole pattern of ascending warm and descending cold air masses (thermal circulation) around the transitioning storm (Klein et al., 2000) and the possible cyclonal flow towards a baroclinic zone trigger baroclinic conversion of eddy available potential into eddy kinetic energy. Cross-contour (ageostrophic) flow from higher towards lower geopotential heights in the outflow exports energy from the transitioning cyclone into the mid-latitude flow (Harr et al., 2000). All in all, the examination of the forecast scenarios for Choi-Wan, once more, highlights the crucial role of the phasing between the transitioning storm and the mid-latitude wave pattern. In all four cases, there already exists a moderately amplified wave train in the extratropics as Choi-Wan undergoes recurvature and starts to interact with the mid-latitude flow. The impact of Choi-Wan on the extratropical wave pattern then depends strongly on the location of the transitioning storm relative to a trough-ridge couplet over Japan and a preexisting extratropical cyclone southeast of Kamchatka. However, the rather short duration Choi-Wan's impact on the extratropical circulation in some scenarios also indicates that there might exist a rather short period of time during which the mid-latitude flow is sensitive to an interaction with a transitioning tropical cyclone. After that time, the extratropical wave pattern is less sensitive and the impact of the tropical cyclone decreases. As Choi-Wan moves into a favouring position ahead of the trough over Japan (scenario III), the outflow of the transitioning storm, together with strong baroclinic conversion due to the thermal circulation and the enhanced northeastwards flow ahead of storm towards a baroclinic zone directly support the further amplification of the adjacent downstream ridge. This downstream support continues as Choi-Wan merges with the extratropical cyclone and the merger reintensifies strongly. As Typhoon Nabi in the study of Harr and Dea (2009), Choi-Wan acts as a strong additional source of eddy kinetic energy for downstream baroclinic development, which reaches into the eastern North Pacific and towards North America. This development has the potential for a ongoing reintensification of Choi-Wan and downstream cyclogenesis, but as the transformed storm separates from the mid-latitude circulation, its further downstream support and reintensification is prohibited. The synoptic set-up and the flux of eddy kinetic energy during the early stages of ET is related in another scenario as well (scenario I), but the impact of Choi-Wan on the downstream mid-latitude flow ceased to exist already during the transformation. In this case, the distance between Choi-Wan and the preexisting extratropical cyclone and its associated trough is larger than in the former scenario and Choi-Wan is rather located to the rear of the mid-latitude trough. During early stages of ET, Choi-Wan supports the further intensification of the extratropical cyclone, which then separates from Choi-Wan and traverses into the eastern North Pacific. The downstream impact of Choi-Wan is then disabled, as the storm becomes more and more embedded in the mid-latitude ridge. The related onset of the development in scenario I and III indicates that the reintensification of Choi-Wan also depends on the development of the extratropical cyclone. If the extratropical cyclone and the associated trough separate from Choi-Wan, its reintensification is apparently prohibited. In contrast to the former scenarios, a strong downstream impact of Choi-Wan during its ET is hampered, if the transitioning storm moves into a highly amplified mid-latitude ridge (scenario

IV). Only as the remnants of Choi-Wan merged with an upstream extratropical cyclone and the dominant ridge is eroded, the impact of the transitioned tropical cyclone on downstream regions increased. Intense baroclinic conversion in the merged system, together with strong downstream dispersion of K_e then lead to a highly amplified wave pattern in the central and eastern North Pacific. Although Choi-Wan is rather strong at the onset of its ET, this case in parts resemble the ET of Typhoon Banyan, examined by Harr and Dea (2009). The downstream impact of Typhoon Banyan was onset after the transitioning storm strongly reintensified as an extratropical system in an weak mid-latitude flow pattern with reduced baroclinicity. The remaining scenario (scenario II) highlights that Choi-Wan also can have no strong impact on the mid-latitude flow. The already weak downstream dispersion of K_e emanating from Choi-Wan during the onset of ET is markedly reduced shortly after the extratropical cyclone and the associated downstream trough move further eastward. However, it is the energy flux from upstream regions through the ridge northwest of Choi-Wan that support the strong intensification of the extratropical cyclone in the eastern North Pacific. This development clearly depends on the mid-latitude circulation itself, without a significant contribution from the transitioning storm.

The different forecast scenarios for the ET of Typhoon Choi-Wan resemble the large variability in downstream response that was also found by Harr and Dea (2009) during their examination of analysis data for four transitioning tropical cyclones in the western North Pacific. However, in our case, the distinct development scenarios do not arise from a set of analysis data for different storms, but arise due to the forecast uncertainty in numerical weather prediction during an ET event. Uncertainties in the representation of the transitioning storm and the mid-latitude flow lead to completely different development scenarios in one ensemble forecast. Parts of these uncertainties result from errors in the initial conditions during the onset of the model and parts result from an inaccurate representation of the underlying physical processes during the model run. In this study, the different mechanisms during the ET causing the impact of the storm on the mid-latitude flow were further examined. A correct representation of the phasing between the transitioning storm and the mid-latitude flow is important to adequately predict the flux of energy from the transitioning cyclone into the extratropical flow and hence to resemble the impact of the tropical cyclone on the mid-latitude wave pattern. Furthermore, it is also important to represent the thermal circulation during the ET process, which triggers the baroclinic conversion of eddy available potential into eddy kinetic energy and hence provide a source for downstream dispersion of K_e.

6 Conclusions and Outlook

The extratropical transition of tropical cyclones exhibits the potential to cause high impact weather events, even in regions far downstream of the ET event. At the same time, the complex physical and dynamical processes during the interaction between the mid-latitude flow and the transitioning tropical cyclone, together with the sensitivity to the phasing between the storm and the extratropical wave pattern, often reduce the reliability of forecasts. A reduction or at least an appropriate consideration of the forecast uncertainty during an ET provides the basis for diminishing the possible impact on downstream regions. This necessitates a further examination of the processes involved and the development of new approaches for how to address the reduction of predictability during operational forecasts. The present study considered parts of both aspects by first examining forecast variability during ET events in a newly established multimodel ensemble system, and second by investigating the energy budget in ensemble forecasts during the interaction between the tropical cyclone and the mid-latitude flow to learn more about the processes involved.

In the first part of the study the potential benefit of a new multimodel ensemble approach in assessing forecast variability during ET events is explored. By applying an EOF- and subsequent cluster analysis, the variability in forecast synoptic patterns associated with ET events in the THOR-PEX Interactive Grand Global Ensemble (TIGGE) was examined. In addition to the common representation of initial uncertainty in ensemble forecasts, this new multimodel ensemble approach considers forecast variability arising from uncertainties in the model formulation itself, as it combines forecasts from several operationally running ensemble prediction systems. The focus was put on characterising the behaviour of the multimodel ensemble in forecasting the ET of ten tropical cyclones in 2008. To assess questions concerning the performance of the multimodel approach compared to a single ensemble forecast, the study was based on the examination of three different data sets: TIGGE as the whole multimodel ensemble, the ECMWF as a single ensemble system and TIGGE without the ECMWF EPS (TI-EC) as a reduced multimodel ensemble. Questions to be addressed during the study were whether the combination of several individual EPS in TIGGE provides increased representation of forecast variability and thus more possible development scenarios compared to the single model, and how the individual ensembles contribute to the forecast variability.

A first examination of forecast variability showed that the ET cases considered had a different impact on the forecast uncertainties in the individual ensemble systems in most of the forecasts under investigation. The application of an EOF analysis then allowed those features of the mid-latitude flow pattern to be identified that exhibit the primary differences among the ensemble members within the three data sets. Overall, the identified variability patterns corroborated the findings from Anwender et al. (2008), in which the forecast variability during ET events is mainly linked to the representation of the amplification of the wave pattern (amplitude pattern) and its longitudinal position (shift pattern). The variability patterns differed between the three data sets. In some cases, the two dominant variability patterns were related in the TIGGE and the TI-EC data sets but different in the ECMWF EPS, indicating a minor contribution of the ECMWF EPS to the variability in the multimodel data set. In other cases, the ECMWF EPS affected the second strongest variability in the TIGGE data set, as its exclusion caused differences in the variability patterns. In one case related dominant variability patterns were found in all three data sets. The subsequent cluster analysis indicated that the individual ensemble systems had different contributions to the clusters extracted and thus to the different possible development scenarios. While some ensemble systems (ECMWF and Japanese EPS) contributed to nearly all scenarios, other systems (Brazilian and Australian EPS) had a contribution to only one or two of the scenarios. This is in agreement with the different representations of forecast variability in the individual ensemble members. The levels of variability achieved in the different EPS seemed to be independent of the initialisation method. The ECMWF EPS was found to often dominate one development scenario, which was missing in the TIGGE data set when the ECMWF EPS was excluded. Overall, the multimodel approach spanned the broadest bandwidth of possible development scenarios, broader than the ECMWF EPS alone. However, the ECMWF EPS, known as a well performing EPS, was necessary to obtain the full scope of possible synoptic developments.

In general, ensemble prediction systems should resemble the real uncertainty about the future atmospheric state. Hence, they should indicate low variability if the development has a high probability of occurrence, while large variability should occur, when the future synoptic situation is uncertain. In other words, the sampled probability density function (pdf) of the ensemble forecast should be related to the pdf of the real atmospheric development. An estimation of the real pdf requires a statistical evaluation of large number of closely related developments with distinct outcomes. The ten cases considered in this study do not meet these demands. Therefore, we cannot address the question as to whether the increased forecast variability in the TIGGE data base compared with the variability in the single ECMWF EPS was justified. However, it could be shown that the verifying analysis would have not captured in the ECMWF EPS in some cases, but was in the scope of the TIGGE data set.

The analysis technique presented here, which was originally used for ET by Harr et al. (2008), performed well during our application and led to a reasonable aggregation of the information contained in the ensemble forecasts under examination. By comparing our method with another clustering approach, the assets and drawbacks of both approaches shall be further examined in the near future. The EOF- and cluster method might also have the potential to be further adapted for operational usage, allowing probability information during periods of high forecast uncertainty to be better accounted for. The clusters provide information about the main possible development scenarios and, at the same time, the number of members contributing to these development scenarios indicate how probable they are to occur. This method was already employed during the T-PARC field campaign to give an indication for possible ET events. In an operational application, the clusters and their probability might offer the forecaster a more condensed overview about the content of the ensemble forecasts instead of a manual comparison of individual ensemble members or considering only the ensemble mean. On the other hand, an operational application of the method also requires further automatisation of the process, especially for the definition of the cluster numbers considered. Another possibility provided by the methodology of analysing TIGGE data presented here is to use the main development scenarios as initial and boundary conditions for high resolution regional ensembles and thus quantify predictability on smaller scales (e.g. associated with convection) in more detail. This preselection might allow specification of a broad and reasonable range of variability in regional ensembles even with a limited number of members. Finally, this study points to the utility of the TIGGE data set and to the need for continued analyses of the information contained in combinations of EPS from various operational systems (Keller et al., 2011).

The second part of the present study focused on a closer examination of the interaction between the transitioning cyclone and the mid-latitude flow pattern. Therefore, several development scenarios from the ECMWF EPS for the ET of Hurricane Hanna (2008) and Typhoon Choi-Wan (2009) were analysed by investigating their eddy kinetic energy budgets during the ET process. However, this first necessitated ensuring that the vertical output resolution of the operational ECMWF EPS was sufficient to capture the distinct budget terms adequately. The development scenarios were then conducted by applying the EOF- and cluster analysis on the 500 hPa geopotential height field and the vertically averaged eddy kinetic energy of an ECMWF ensemble forecasts, respectively initialised prior to the ET of the two transitioning storms. In the case of Hanna, the two clustering approaches, although they partially differed in the standard deviation patterns, resulted in two dominant possible development scenarios for the ET of Hanna. On the other hand, four distinct forecast scenarios were found for the ET of Choi-Wan. These scenarios occurred in both clustering solutions.

For both storms, the examination of the different K_e budgets then provided insight into the different processes that were involved in the extratropical transition. In the case of Hanna, the differences in the two scenarios could be linked to a distinct contribution of baroclinic conversion from available potential into kinetic energy during the ET process. As Hanna dispersed K_e towards the mid-latitude flow during the early stages of its ET, baroclinic conversion nearly ceased to exist as Hanna moved off the coast of the North American mainland, ahead of a weak trough. Baroclinic conversion, as well as energy fluxes then dropped completely and the downstream wave pattern did not amplify further. In the contrasting scenario, baroclinic conversion was onset later during the ET process and continued to exist as Hanna was located ahead of a short downstream trough. The ongoing generation of K_e then supported the further amplification of the wave pattern and preconditioned the reintensification of Hanna. Baroclinic conversion also contributed to the development of four different scenarios for the ET of Choi-Wan, but these scenarios also highlighted the crucial role of the phasing between the transitioning storm and the mid-latitude wave pattern. In two of the four scenarios, Choi-Wan clearly enhanced the K_e of the mid-latitude flow pattern during the early ET phase. As the storm was then located ahead of a trough and merged with a preexisting extratropical cyclone, its strong influence on the amplification of the mid-latitude wave pattern continued until the system cut off in the central North Pacific. In contrast, as the extratropical cyclone and the associated trough started to separate gradually from the transitioning Choi-Wan in the other scenario, its strong influence on the mid-latitude flow pattern decreased significantly. In a third case, Choi-Wan transitioned into a strongly amplified ridge in the western North Pacific. This ridge suppressed a strong reintensification of Choi-Wan. However, the transitioning storm dispersed energy into the downstream region where the extratropical cyclone intensified. This support was further enhanced and led to a strong amplification of the mid-latitude wave pattern as Choi-Wan merged with another extratropical cyclone west of Kamchatka. In the fourth scenario Choi-Wan had an only minor impact on the mid-latitude flow and the development of a strongly amplified wave pattern was mainly triggered from the upstream mid-latitude regions. Upstream support in the other cases, in contrast, was rather weak.

The strong impact of baroclinic conversion during the ET of Hanna coincides with other studies concerning ET processes in general and Hanna in particular. Grams et al. (2011) could highlight the significant impact on the amplification of an adjacent mid-latitude ridge during the ET of Hurricane Hanna. In his investigation of Hurricane Hazel (1954), Palmén (1958) already mentioned the importance of baroclinic processes during the interaction between the mid-latitude flow and the transitioning tropical cyclone. In the case of Choi-Wan, the importance of the phasing between the transitioning storm and the mid-latitude wave pattern was emphasised, but the

ET process seemed also to depend on the interaction between Choi-Wan and the pre-existing extratropical cyclone. Choi-Wan reintensified strongly and had a clear impact on the mid-latitude flow, as it merged with this extratropical feature. As the extratropical cyclone started to separate from Choi-Wan, its strong impact decreased. Furthermore, the mid-latitude flow seems to be sensitive to an interaction with the transitioning tropical cyclone only during a short period of time. The development scenarios during the ET of Choi-Wan clearly highlight a part of the different roles, the transitioning storm can play during its interaction with the mid-latitude flow. Harr and Dea (2009) could also identify distinct impacts of the transitioning storm on the mid-latitude flow, but as their examinations were based on analysis data, they could only hypothesise how a distinct behaviour of the transitioning storm and its interaction with the mid-latitude flow would affect the downstream development. In this work, the different possible impacts for one specific storm could be examined.

The analysis of the EKE-budget for the two transitioning storm indicated different aspects that influence the outcome of the transitioning storm, but also raised a series of subsequent questions requiring further examination. In the case of Choi-Wan it would be interesting to investigate whether the extratropical cyclone, which was crucial for the ET of Choi-Wan, was actually triggered by the transitioning storm itself, meaning that the storm might have aided its own reintensification. This question could be answered by conducting model runs with high resolution for the amplification generation of this cyclone. Further insight into the interaction between the processes involved could also be gained by sensitivity studies using the full ensemble forecasts. Thereby, the correlation between two forecast variables at different times, i.e. the amplification of the downstream trough at a later forecast time and the energy flux emanating from Choi-Wan at an earlier forecast time, can provide further information about how the different features of the synoptic flow pattern depend on each other.

By investigating different approaches, the present study provides an additional piece to the puzzle of assessing and further understanding the impact of an extratropical transition of a tropical cyclone on the predictability for downstream regions. Although our understanding of the ET process itself and its impact on predictability could be further deepened during the last couple of years, further research is still required to solve the remaining questions and thus to better consider the increased forecast uncertainty during ET events.

Acknowledgment

This work was conducted in the frame of the German Research Group PANDOWAE (Predictability and Dynamics of the Atlantic-European Centre) at the Karlsruhe Institute of Technology (KIT). With the submission of this thesis, my time as a PANDOWAE PhD student is also drawing to a close. The recent three years were marked by exciting and formative experiences on my way to becoming a member of the research community, which affected my working and also my private life. There are some people who made this time to a particular one and whose help enabled the thesis in this manner at all.

Most of all I want to thank my supervisor Prof. Sarah Jones. She enabled me to become a member of PANDOWAE and to conduct this thesis in such an unique framework. Her continuous support, experience and commitment to our research field encouraged me every time anew to find my way through this work. In precious and inspiring discussions she animated me to explore the various aspects of my work in particular, and of Meteorology and science in general. She strongly encouraged collaborations with international colleagues and gave me the great opportunity to some experiences in international cooperations. Sarah, thank you for all your support, encouragement and for spreading of so much positive energy and thank you for being not only a supervisor but also being a friend. I am looking forward to the Badische Meile 2012!

I also want to thank Prof. Christoph Kottmeier for accepting and taking time to co-supervise this work and showing his interest in valuable comments.

Furthermore, I am also very grateful to thank Prof. Pat Harr. He kindly provided most of the analysis methods employed in this study. Prof. Harr also invited me to visit the Naval Postgraduate School, Monterey and attended to me as a warm and genial host. He was always interested in my work and I very much appreciate his precious comments, ideas and suggestions based on his great experience in our research field.

Prof. Jenni Evans provided me the possibility to visit the Penn State University and to work on joint study. I want to thank her for enabling this inspiring visit in State College and for the valuable and helpful discussions about my work. Her great experience helped me a lot to learn more about and to further explore the different aspects of the cluster analysis. Thank you also for being a great host.

I am also very grateful to Dr. Chris Davis, who gave me the possibility to visit NCAR and had many valuable suggestions and ideas. I owe Prof. Edmund Chang, Stony Brook University, many thanks for his great and patient support concerning the eddy kinetic energy analysis code and for his valuable advices during the interpretation. Precious and encouraging discussions with Dr. Carolyn Reynolds and Dr. Jim Doyle from the Naval Research Laboratory, Monterey, and Prof. Francesca Chiaromonte from Penn State University were very helpful, especially concerning the analysis of the TIGGE data.

Many thanks also go to the THORPEX consortium and the contributing weather services for providing the TIGGE data base and in particular to the ECMWF for maintaining the TIGGE archive and for furthermore providing the ensemble forecasts employed in the eddy kinetic energy analysis. I also want to express my gratitude to the German Research Council, who provided funding for the research unit PANDOWAE. Special thanks go to the Karlsruhe House of Young Scientists (KHYS), whose travel funding enabled me a three month stay abroad.

All members of PANDOWAE made this research group to an unique working environment. I want to thank all of you for the great and inspiring workshops and discussions. I am looking forward to the second phase!

I especially want to thank Christian Grams and Simon Lang, my "PhD brothers". Thank you for your encouragement, support and friendship during all stages of the graduation. It was a great pleasure for me to graduate together with both of you. We made it! Many thanks also go to all of my other colleagues in our institute who made our floor to an exceptional and enjoyable working environment. In particular I want to thank Doris Anwender, Leonhard Scheck, Juliane Schwendike and Hans Schipper for their friendship and support during all the time.

All of this would have not been possible without the absolute and everlasting support by my parents Gaby Keller and Ulrich Kretschmer-Keller, in every way. Thank you for being there, every time, during all ups and downs of (PhD) life, for your trust in me and your perpetual encouragement! I especially want to thank my mother for taking care of my horse during all the stressful times.

My very special thanks go to my partner Wolfgang Woiwode. Thank you for your invaluable support and encouragement all the time, for your love and comfort and for showing me the light at the end of the tunnel when I could not see it!

Bibliography

Agustí-Panareda, A., Gray, S., Craig, G., and Thorncroft, C. (2005). The Extratropical Transition of Tropical Cyclone Lili (1996) and its Crucial Contribution to a Moderate Extratropical Development. *Mon. Weather Rev.*, 133:1562–1573.

Anderberg, M. R. (1973). *Cluster Analysis for Applications*. Academic Press.

Anwender, D. (2007). *Extratropical Transition in the Ensemble Prediction System of the ECMWF: Case Studies and Experiments*. PhD thesis, University of Karlsruhe.

Anwender, D., Harr, P., and Jones, S. (2008). Predictability Associated with the Downstream Impacts of the Extratropical Transition of Tropical Cyclones: Case Studies. *Mon. Weather Rev.*, 136:3226–3247.

Arnott, J. M., Evans, J. L., and Chiaromonte, F. (2004). Characterization of Extratropical Transition Using Cluster Analysis. *Mon. Weather Rev.*, 132:2916–2937.

Atallah, E. H. and Bosart, L. F. (2003). The Extratropical Transition and Precipitation Distribution of Hurricane Floyd (1999). *Mon. Weather Rev.*, 131:1063–1081.

Berner, J., Shutts, G., Leutbecher, M., and Palmer, T. (2009). A Spectral Stochastic Kinetic Energy Backscatter Scheme and its Impact on Flow-

Dependent Predictability in the ECMWF Ensemble Prediction System. *J. Atmos. Sci.*, 66:603–626.

Bezdek, J. C. (1981). *Pattern Recognition with Fuzzy Objective Function Algorithms*. Plenum Press.

Bezdek, J. C., Ehrlich, R., and Full, W. (1984). FCM: The Fuzzy C-Means Clustering Algorithm. *Computers & Geosciences*, 10:191–203.

Bishop, C. H., Etherton, B. J., and Majumdar, S. J. (2001). Adaptive Sampling with the Ensemble Transform Kalman Filter. Part I: Theoretical Aspects. *Mon. Weather Rev.*, 129:420–436.

Björnsson, H. and Venegas, S. A. (1997). A Manual for EOF and SVD Analyses of Climatic Data. CCGCR Rep. 97-1, Centre for Climate and Global Change Research, McGill University.

Blake, E. and Gibney, E. (2011). *The Deadliest, Costliest, and Most Intense United States Tropical Cyclones From 1851 to 2010 (and Other Frequently Requested Hurricane Facts)*. National Weather Service, National Hurricane Center.

Böker, F. (2005). Multivariate Verfahren. Technical report, Georg-August-Universität Göttingen.

Bougeault, P., Toth, Z., Bishop, C., Brown, B., Burridge, D., Chen, D., Ebert, B., Fuentes, M., Hamill, T. M., Mylne, K., Nicolau, J., Paccagnella, T., Park, Y.-Y., Parsons, D., Raoult, B., Schuster, D., Dias, P. S., Swinbank, R., Takeuchi, Y., Tennant, W., Wilson, L., and Worley, S. (2010). The THORPEX Interactive Grand Global Ensemble. *Bull. Am. Meteorol. Soc.*, 91:1059–1072.

Bouttier, F. and Courtier, P. (1999). *Data Assimilation Concepts and Methods*. ECMWF, Shinfield Park, Reading UK, Shinfield Park, Reading UK.

Briegel, L. and Frank, W. (1997). Large-Scale Influences on Tropical Cyclogenesis in the Western North Pacific. *Mon. Weather Rev.*, 125:1397–1413.

Brown, D. and Kimberlain, T. (2008). *Tropical Cyclone Report, Hurricane Hanna, (AL082008), 28 August - 7 September 2008*. National Hurricane Center.

Browning, K., Vaughan, G., and Panagi, P. (1998). Analysis of an Ex-Tropical Cyclone After Reintensifying as a Warm-Core Extratropical Cyclone . *Quart. J. Roy. Meteorol. Soc.*, 124:2329–2356.

Buizza, R. (1994). Sensitivity of Optimal Unstable Structures. *Quart. J. Roy. Meteorol. Soc.*, 120:429–451.

Buizza, R. (2003). Weather Prediction: Ensemble Prediction. In Holton, J., Pyle, J., and Curry, J., editors, *Encyclopedia of Atmospheric Sciences*. Academic Press.

Buizza, R. (2006). The ECMWF Ensemble Prediction System. In Palmer, T. and Hagedorn, R., editors, *Predictability of Weather and Climate*, chapter 17, pages 459–488. Cambridge University Press.

Buizza, R. and Palmer, T. (1995). The Singular-Vector Structure of the Atmospheric General Circulation. *J. Atmos. Sci.*, 52:1434–1456.

Chang, E. and Orlanski, I. (1993). On the Dynamics of a Storm Track. *J. Atmos. Sci.*, 50:999 – 1015.

Chang, E. K. (1993). Downstream Development of Baroclinic Waves as Inferred from Regression Analysis. *J. Atmos. Sci.*, 50:2038–2053.

Chang, E. K. (2000). Wave Packets and Life Cycles of Troughs in the Upper Troposphere: Examples from the Southern Hemisphere Summer Season of 1984/1985. *Mon. Weather Rev.*, 128:25–50.

Charney, J. (1947). The Dynamics of Long Waves in a Baroclinic Westerly Current. *J. Meteorol.*, 4:135–162.

Cheng, X. and Wallace, J. M. (1991). Cluster Analysis of the Northern Hemisphere Wintertime 500-hPa Height Field: Spatial Patterns. *J. Atmos. Sci.*, 50:2674–2695.

Chien, H. and Smith, P. (1977). Synoptic and Kinetic Energy Analysis of Hurricane Camille (1969) During Transition Across the Southeastern United States. *Mon. Weather Rev.*, 105:67–77.

Christiansen, B. (2007). Atmospheric Circulation Regimes: Can Cluster Analysis Provide the Number? *J. Clim.*, 20:2229–2250.

Cooper, G. and Falvey, R. (2008). *Annual Tropical Cyclone Report 2008*. Joint Typhoon Warning Centre.

Daley, R. (1991). *Atmospheric Data Analysis*. Cambridge University Press.

Danielson, R. E., Gyakum, J. R., and Straub, D. N. (2004). Downstream Baroclinic Development Among forty-one Cold-Season Eastern North Pacific Cyclones. *Atmos. Ocean*, 43:235–250.

Davis, C. A. and Emanuel, K. (1991). Potential Vorticity Diagnostics of Cyclogenesis. *Mon. Weather Rev.*, 119:1929–1953.

Decker, S. G. and Martin, J. E. (2005). A Local Energetics Analysis of the Life Cycle Differences Between Consecutive, Explosively Deepening Continental Cyclones. *Mon. Weather Rev.*, 133:295–316.

DiMego, G. and Bosart, L. (1982). The Transformation of Tropical Storm Agnes into an Extratropical Cyclone. Part II: Moisture, Vortictiy and Kinetic Energy Budgets. *Mon. Weather Rev.*, 110:412–433.

Eady, E. (1949). Long Waves and Cyclone Waves. *Tellus*, 1:33–52.

Eady, E. (1951). The Quantitative Theory of Cyclone Development. In Malone, T., editor, *Compendium of Meteorology*. American Meteorological Society.

Emanuel, K. (1986). An Air-Sea Interaction Theory for Tropical Cyclones, Part I: Steady-State Maintenance. *J. Atmos. Sci.*, 42:1279–1293.

Emanuel, K. (1991). The Theory of Hurricanes. *Annu. Rev. Fluid Mech.*, 23:179–196.

Emanuel, K. (2003). Tropical Cyclones. *Annu. Rev. Earth Planet. Sci.*, 31:75–104.

Epstein, E. (1969). Stochastic Dynamic Perturbation. *Tellus*, 21:739–759.

Evans, J., Arnott, J., and Chiaromonte, F. (2006). Evaluation of Operational Model Cyclone Structure Forecasts During Extratropical Transition. *Mon. Weather Rev.*, 134:3054–3072.

Evans, J. L. and Hart, R. E. (2003). Objective Indicators of the Life Cycle Evolution of Extratropical Transition for Atlantic Tropical Cyclones. *Mon. Weather Rev.*, 131:909–925.

Farrell, B. (1982). The Initial Growth of Disturbances in a Baroclinic Flow. *J. Atmos. Sci.*, 39:1663–1686.

Farrell, B. (1984). Modal and Nonmodal Baroclinic Waves. *J. Atmos. Sci.*, 41:668–673.

Farrell, B. (1985). Transient Growth of Damped Baroclinic Waves. *J. Atmos. Sci.*, 42:2718–2727.

Farrell, B. (1989). Optimal Excitation of Baroclinic Waves. *J. Atmos. Sci.*, 46:1193–1206.

Förster, A. (2011). *Structural Characteristics of the Core Region of T-PARC Typhoon Sinlaku in a Vertically Sheared Environment.* Master's thesis, Karlsruhe Institute of Technology (KIT), Karlsruhe, Germany.

Froude, L. S. R. (2010). TIGGE: Comparison of the Prediction of Northern Hemisphere Extratropical Cyclones by Different Ensemble Prediction Systems. *Wea. Forecasting*, 25:819–836.

Glatt, I., Dörnbrack, A., Jones, S. C., Keller, J. H., Martius, O., Müller, A., Peters, D. H., and Wirth, V. (2011). Utility of Hovmöller Diagrams to Diagnose Rossby Wave Trains. *Tellus A*, 63:991–1006.

Grams, C. (2011). *Quantification of the Downstream Impact of the Extratropical Transition of Typhoon Jangmi and other Case Studies.* PhD thesis, Karlsruhe Institute of Technology.

Grams, C. M., Wernli, H., Böttcher, M., Čampa, J., Corsmeier, U., Jones, S. C., Keller, J. H., Lenz, C.-J., and Wiegand, L. (2011). The Key Role of Diabatic Processes in Modifying the Upper-Tropospheric Wave Guide: a North Atlantic Case Study. *Quart. J. Roy. Meteorol. Soc.*, 137:2174–2193.

Gray, W. (1968). Global View on the Origin of Tropical Disturbances and Storms. *Mon. Weather Rev.*, 96:669–700.

Grazzini, F. and van der Grijn, G. (2003). Central European Floods During Summer 2002. *ECMWF Newsletter*, 96:18–28.

Haltiner, G. J. and Williams, R. (1979). *Numerical Prediction and Dynamic Meteorology.* Wiley & Sons, 2nd edition.

Hamill, T. (2006). Ensemble-Based Atmospheric Data Assimilation. In Palmer, T. and Hagedorn, R., editors, *Predictability of Weather and Climate*, chapter 6, pages 124–156. Cambridge University Press.

Hannachi, A. (2004). *A Primer for EOF Analysis of Climate Data.* Department of Meteorology, University of Reading.

Hannachi, A., Jolliffe, I., and Stephenson, D. B. (2007). Empirical Orthogonal Functions and Related Techniques in Atmospheric Science: A Review. *Int. J. Climatology*, 27:1119–1152.

Harr, P. (2010). The Extratropical Transition of Tropical Cyclones. In Chan, J. and Kepert, J., editors, *Global Perspectives on Tropical Cyclones, From Science to Mitigation*, volume 4 of *World Scientific Series on Asia-Pacific Weather and Climate*, chapter 5, pages 149–176. World Scientific, 2nd edition.

Harr, P. A., Anwender, D., and Jones, S. C. (2008). Predictability Associated with the Downstream Impacts of the Extratropical Transition of Tropical Cyclones: Methodology and a Case Study of Typhoon Nabi (2005). *Mon. Weather Rev.*, 136:3205–3225.

Harr, P. A. and Dea, J. M. (2009). Downstream Development Associated with the Extratropical Transition of Tropical Cyclones over the Western North Pacific. *Mon. Weather Rev.*, 137:1295–1319.

Harr, P. A. and Elsberry, R. L. (1995). Large-Scale Circulation Variability over the Tropical Western North Pacific. Part II: Persistence and Transition Characteristics. *Mon. Weather Rev.*, 123:1247–1268.

Harr, P. A. and Elsberry, R. L. (2000). Extratropical Transition of Tropical Cyclones over the Western North Pacific. PartI: Evolution of Structural Characteristics During the Transition Process. *Mon. Weather Rev.*, 128:2613–2633.

Harr, P. A., Elsberry, R. L., and Hogan, T. F. (2000). Extratropical Transition of Tropical Cyclones over the western North Pacific. PartII: The Impact of Midlatitude Circulation Characteristics. *Mon. Weather Rev.*, 128:2634–2653.

Hart, R. (2003). A Cyclone Phase Space Derived from Thermal Wind and Thermal Asymmetry. *Mon. Weather Rev.*, 131:585–616.

Hart, R. and Evans, J. (2001). A Climatology of the Extratropical Transition of Atlantic Tropical Cyclones. *J. Climate*, 14:546–564.

Holton, J. R. (2004). *An Introduction to Dynamic Meteorology.* Elsevier Academic Press.

Hoskins, B. and Valdes, P. (1990). On the Existence of Storm-Tracks. *J. Atmos. Sci.*, 47:1854–1864.

Hoskins, B. J., McIntyre, M., and Robertson, A. (1985). On the Use and Significance of Isentropic Potential Vorticity Maps. *Quart. J. Roy. Meteorol. Soc.*, 111:877–946.

Houtekamer, P., Lefaivre, L. Derome, J., Ritchie, H., and Mitchell, H. (1996). A System Simulation Approach to Ensemble Prediction. *Mon. Weather Rev.*, 124:1225–1242.

Houtekamer, P. and Mitchell, H. (1998). Data Assimilation Using an Ensemble Kalman Filter Technique. *Mon. Weather Rev.*, 126:796.811.

Houtekamer, P. and Mitchell, H. (2005). Ensemble Kalman Filtering. *Quart. J. Roy. Meteorol. Soc.*, 131:3269–3289.

Hovmöller, E. (1949). The Trough-and-Ridge Diagram. *Tellus*, 1:62–66.

JMA (2008). *Annual Report on the Activities of the RSMC Tokyo - Typhoon Center 2008.* Japan Meteorological Agency.

JMA (2009). *Annual Report on the Activities of the RSMC Tokyo - Typhoon Center 2009.* Japan Meteorological Agency.

Johnson, C. and Swinbank, R. (2009). Medium-Range Multimodel Ensemble Combination and Calibration. *Quart. J. Roy. Meteorol. Soc.*, 135:777–794.

Jones, S. C., Harr, P. A., Abraham, J., Bosart, L. F., Bowyer, P. J., Evans, J. L., Hanley, D. E., Hanstrum, B. N., Hart, R. E., Laurette, F., Sinclair, M. R., Smith, R. K., and Thorncroft, C. (2003). The Extratropical Transition of Tropical Cyclones: Forecast Challenges, Current Understanding, and Future Directions. *Wea. Forecasting*, 18:1052–1092.

Kalkstein, L. S., Tan, G., and Skindlov, J. A. (1987). An Evaluation of Three Clustering Procedures for Use in Synoptic Climatological Classification. *J. Clim. a. Appl. Meteorol.*, 26:717–730.

Kalman, R. (1960). A New Approach to Linear Filtering and Prediction Problems. *Trans. ASME. Ser. D, J. Basic Engineering*, 82:35–45.

Kalnay, E. (2003). *Atmospheric Modeling, Data Assimilation and Predictability*. Cambridge University Press.

Kalnay, E. (2009). Introduction to Ensemble Kalman Filter. In *Summer Colloquium on Data Assimilation*. Joint Center for Satellite Data Assimilation.

Keller, J. H., Jones, S. C., Evans, J. L., and Harr, P. (2011). Characteristics of the TIGGE Multimodel Ensemble Prediction System in Representing Forecast Variability Associated with Extratropical Transition. *Geophys. Res. Lett.*, 38:L12802.

Ketchen, D. J. and Shook, C. L. (1996). The Application of Cluster Analysis in Strategic Management Research: An Analysis and Critique. *Strat. Management J.*, 17:441–458.

Klein, P., Harr, P., and Elsberry, R. (2002). Extratropical Transition of Western North Pacific Cyclones: Midlatitude and Tropical Cyclone Contributions to Reintensification. *Mon. Weather Rev.*, 130:2240–2259.

Klein, P. M., Harr, P. A., and Elsberry, R. L. (2000). Extratropical Transition of Western North Pacific Tropical Cyclones: An Overview and Conceptual Model of the Transformation Stage. *Wea. Forecasting*, 15:373–396.

Kornegay, F. and Vincent, D. (1976). Kinetic Energy Budget Analysis During Interaction of Tropical Storm Candy (1968) with an Extratropical Frontal System. *Mon. Weather Rev.*, 104:849–859.

Kung, E. (1966). Kinetic Energy Generation and Dissipation in the Large-Scale Atmospheric Circulation. *Mon. Weather Rev.*, 94:67–82.

Kung, E. (1967). Diurnal and Long-Term Variations of the Kinetic Energy Generation and Dissipation for a Five-Year Period. *Mon. Weather Rev.*, 95:593–606.

Kung, E. (1977). Energy Sources in Middle-Latitude Synoptic-Scale Disturbances. *J. Atmos. Sci.*, 34:1352–1365.

Kung, E. and Baker, W. (1975). Energy Transformations in Middle-Latitude Disturbances. *Quart. J. Roy. Meteorol. Soc.*, 101:793–815.

Kutzbach, J. E. (1967). Empirical Eigenvectors of Sea-Level Pressure, Surface Temperature and Precipitation Complexes over North America. *J. Appl. Meteorol.*, 6:791–802.

Lang, S. T. K. (2011). *Perturbation Dynamics and Impact of Different Perturbation Methods in Tropical Cyclone Ensemble Forecasting*. PhD thesis, Karlsruhe Institute of Technology.

Leith, C. (1974). Theoretical Skill of Monte Carlo Forecasts. *Mon. Weather Rev.*, 102:409–418.

Leutbecher, M. and Palmer, T. (2008). Ensemble Forecasting. *J. Comp. Phys.*, 227:3515–3539.

Lewis, J. (2005). Roots of Ensemble Forecasting. *Mon. Weather Rev.*, 133:1856–1885.

Lim, G. H. and Wallace, J. M. (1991). Structure and Evolution of Baroclinic Waves as Inferred from Regression Analysis. *J. Atmos. Sci.*, 48:1718–1732.

Lorenz, E. (1956). Empirical Orthogonal Functions and Statistical Weather Prediction. Statistical Forecasting Project Sci. Rep. No. 1, Dept. of Meteorology, MIT.

Lorenz, E. (1962). The Statistical Prediction of Solutions of Dynamic Equations. Proc. Intern. Symp. Numer. Weather Pred. Tokyo. *J. Meteor. Soc.*, 647:629–635.

Lorenz, E. (1963). Deterministic Nonperiodic Flow. *J. Atmos. Sci.*, 20:130–141.

Lorenz, E. (1982). Atmospheric Predictability Experiments with Large Numerical Models. *Tellus*, 34:505–513.

Lorenz, E. N. (1955). Available Potential Energy and the Maintenance of the General Circulation. *Tellus*, 7:157–167.

Magnusson, L., Leutbecher, M., and Källen, E. (2008). Comparison between Singular Vectors and Breeding Vectors as Initial Perturbations for the ECMWF Ensemble Prediction System. *Mon. Weather Rev.*, 136:4092–4104.

Majumdar, S. J. and Finocchio, P. (2010). On the Ability of Global Ensemble Prediction Systems to Predict Tropical Cyclone Track Probabilities. *Wea. Forecasting*, 25:659–680.

Martius, O., Schwierz, C., and Davies, H. C. (2008). Far-Upstream Precursors of Heavy Precipitation Events on the Alpine South-Side. *Quart. J. Roy. Meteorol. Soc.*, 134:417–428.

Matsueda, M. and Tanaka, H. (2008). Can MCGE Outperform the ECMWF Ensemble? *SOLA*, 4:77–80.

McLay, J. and Martin, J. (2002). Surface Cyclolysis in the North Pacific Ocean. Part III: Composite Local Energetics of Tropospheric-Deep Cyclone Decay Associated with Rapid Surface Cyclolysis. *Mon. Weather Rev.*, 130:2507–2529.

Milligan, G. W. and Cooper, M. C. (1985). An Examination of Procedures for Determining the Number of Clusters in a Data Set. *Psychometrika*, 50:159–179.

Mo, K. and Ghil, M. (1988). Cluster Analysis of Multiple Planetary Flow Regimes. *J. Geophys. Res.*, 93:10927–10952.

Orlanski, I. and Chang, E. K. (1993). Ageostrophic Geopotential Fluxes in Downstream and Upstream Development of Baroclinic Waves. *J. Atmos. Sci.*, 50:212–225.

Orlanski, I. and Katzfey, J. J. (1991). The Life Cycle of a Cyclonic Wave in the Southern Hemisphere: Part I: Eddy Energy Budget. *J. Atmos. Sci.*, 48:1972 – 1998.

Orlanski, I. and Sheldon, J. (1993). A Case of Downstream Baroclinic Development over Western North America. *Mon. Weather Rev.*, 121:2929–2950.

Orlanski, I. and Sheldon, J. (1995). Stages in the Energetics of Baroclinic Systems. *Tellus*, 47 A:605–628.

Ott, E., Hunt, B., Szunyogh, I., Zimin, A., Koestlich, E., Corazza, M., Kalnay, E., Patil, D., and Yorke, J. (2004). A Local Ensemble Kalman filter for Atmospheric Data Assimilation. *Tellus A*, 56:415–428.

Palmén, E. (1948). On the Formation and Structure of Tropical Cyclones. *Geophysica*, 3:27–38.

Palmén, E. (1958). Vertical Circulation and Release of Kinetic Energy During the Development of Hurricane Hazel (1954) into an Extratropical Storm. *Tellus*, 10:1–13.

Palmer, T. (2001). A Nonlinear Dynamical Perspective on Model Error: A Proposal for Non-Local Stochastic-Dynamic Parametrizations in Weather and Climate Prediction Models. *Quart. J. Roy. Meteorol. Soc.*, 127:279–304.

Palmer, T. and Williams, P. (2008). Introduction. Stochastic Physics and Climate Modelling. *Philosophical Transactions of the Royal Society A*, 366:2419–2425.

Pang-Ning Tan, Michael Steinbach, V. K. (2005). *Introduction to Data Mining*. Addison-Wesley.

Park, Y.-Y., Buizza, R., and Leutbecher, M. (2008). TIGGE: Preliminary Results on Comparing and Combining Ensembles. *Quart. J. Roy. Meteorol. Soc.*, 134:2029–2050.

Persson, A. (1998). How Do We Understand the Coriolis Force? *Bull. Am. Meteorol. Soc.*, 79:1373–1385.

Persson, A. and Grazzini, F. (2007). *User Guide to ECMWF Forecast Products*. European Centre for Medium-Range Weather Forecasts, ECMWF Shinfield Park Reading, 4.0 edition.

Pettersen, S. and Smebye, S. (1971). On the Development of Extratropical Storms. *Quart. J. Roy. Meteorol. Soc.*, 97:457–482.

Pichler, H. (1997). *Dynamik der Atmosphäre*. Spektrum Akademischer Verlag.

Plumb, A. R. (1985). On the Three-Dimensional Propagation of Stationary Waves. *J. Atmos. Sci*, 42:217–229.

Quinting, J. (2011). *Structural Characteristics of Typhoon Sinlaku (2008) During its Extratropical Transition: an Observational Study*. Master's thesis, Karlsruhe Institute of Technology (KIT), Karlsruhe, Germany.

Rand, W. M. (1971). Objective Criteria for the Evaluation of Clustering Methods. *J. Am. Stat. Assoc.*, 66(336):846–850.

Reed, R. J. (1990). *Advances in Knowledge and Understanding of Extratropical Cyclones During the Past Quarter Century: An Overview*, chapter 3, pages 27–47. Extratropical Cyclones: The Erik Palmén Memorial Volume. American Meteorological Society.

Riemer, M., Jones, S. C., and Davis, C. A. (2008). The Impact of Extratropical Transition on the Downstream Flow: An Idealized Modelling Study with a Straight Jet. *Q. J. R. Meteorol. Soc.*, 134:69–91.

Ritchie, E. and Elsberry, R. (2003). Simulations of the Extratropical Transition of Tropical Cyclones: Contributions by the Midlatitude Upper-Level Trough to Reintensification. *Mon. Weather Rev.*, 131:2112–2128.

Ritchie, E. A. and Elsberry, R. L. (2007). Simulations of the Extratropical Transition of Tropical Cyclones: Phasing Between the Upper-Level Trough and Tropical Cyclones. *Mon. Weather Rev*, 135:862–876.

Rossby, C. (1940). Planetary Flow Patterns in the Atmosphere. *Quart. J. Roy. Meteorol. Soc.*, 66:68–87.

Sanabia, E. (2010). *The Reintensification of Typhoon Sinlaku*. PhD thesis, Naval Postgraduate School.

Shapiro, M. and Thorpe, A. (2004). *THORPEX International Science Plan*. WWRP/THORPEX No. 2.

Shutts, G. (2005). A Kinetic Energy Backscatter Algorithm for Use in Ensemble Prediction Systems. *Quart. J. Roy. Meteorol. Soc.*, 131:3079–3102.

Shutts, G. (2010). Using Stochastic Physics to Represent Model Error. In *Presentations of the ECMWF NWP-PR Training Course*. ECMWF.

Simmons, A. and Hollingsworth, A. (2002). Some Aspects of the Improvement in Skill of Numerical Weather Prediction. *Quart. J. Roy. Meteorol. Soc.*, 128:647–677.

Thorncroft, C. and Jones, S. (2000). The Extratropical Transitions of Hurricanes Felix and Iris in 1995. *Mon. Weather Rev.*, 128:947–972.

Torn, R. D. and Hakim, G. (2009). Initial Condition Sensitivity of Western Pacific Extratropical Transitions Determined Using Ensemble-Based Sensitivity Analysis. *Mon. Weather Rev.*, 137:3388–3406.

Tory, K. and Frank, W. (2010). Tropical Cyclone Formation. In Chan, J. and Kepert, J., editors, *Global Perspectives on Tropical Cyclones, From Science to Mitigation*, volume 4 of *World Scientific Series on Asia-Pacific Weather and Climate*, chapter 2, pages 55–92. World Scientific, 2nd edition.

Toth, Z. and Kalnay, E. (1993). Ensemble Forecasting at NMC: The Generation of Perturbations. *Bull. Am. Meteorol. Soc.*, 74:2317–2330.

Toth, Z. and Kalnay, E. (1997). Ensemble Forecasting at NCEP and the Breeding Method. *Mon. Weather Rev.*, 125:3297–3319.

Tracton, M. and Kalnay, E. (1993). Ensemble Forecasting at NMC: Practical Aspects. *Wea. Forecasting*, 8:379–398.

Verret, R. (2010). Ensemble Construction. In *EPS Training, MSC*.

von Storch, H. and Navarra, A., editors (1995). *Analysis of Climate Variability: Applications of Statistical Techniques*. Springer.

von Storch, H. and Zwiers, F. W. (1999). *Statistical Analysis in Climate Research*. Cambridge University Press.

Wang, X. and Bishop, C. (2003). A Comparison of Breeding and Ensemble Transform Kalman Filter Ensemble Forecast Schemes. *J. Atmos. Sci.*, 60:1140–1158.

Wilks, D. S. (1995). *Statistical Methods in the Atmospheric Sciences*. Academic Press.

Zadeh, L. (1965). Fuzzy Sets. *Information and Control*, 8:338–353.

Zhang, Z. and Krishnamurti, T. (1999). A Perturbation Method for Hurricane Ensemble Predictions. *Mon. Weather Rev.*, 127:447–469.

Wissenschaftliche Berichte des Instituts für Meteorologie und Klimaforschung des Karlsruher Instituts für Technologie (0179-5619)

Bisher erschienen:

Nr. 1: *Fiedler, F. / Prenosil, T.*
Das MESOKLIP-Experiment. (Mesoskaliges Klimaprogramm im Oberrheintal).
August 1980

Nr. 2: *Tangermann-Dlugi, G.*
Numerische Simulationen atmosphärischer Grenzschichtströmungen über langgestreckten mesoskaligen Hügelketten bei neutraler thermischer Schichtung.
August 1982

Nr. 3: *Witte, N.*
Ein numerisches Modell des Wärmehaushalts fließender Gewässer unter Berücksichtigung thermischer Eingriffe.
Dezember 1982

Nr. 4: *Fiedler, F. / Höschele, K. (Hrsg.)*
Prof. Dr. Max Diem zum 70. Geburtstag.
Februar 1983 (vergriffen)

Nr. 5: *Adrian, G.*
Ein Initialisierungsverfahren für numerische mesoskalige Strömungsmodelle.
Juli 1985

Nr. 6: *Dorwarth, G.*
Numerische Berechnung des Druckwiderstandes typischer Geländeformen.
Januar 1986

Nr. 7: *Vogel, B.; Adrian, G. / Fiedler, F.*
MESOKLIP-Analysen der meteorologischen Beobachtungen von mesoskaligen Phänomenen im Oberrheingraben.
November 1987

Nr. 8: *Hugelmann, C.-P.*
Differenzenverfahren zur Behandlung der Advektion.
Februar 1988

Nr. 9: *Hafner, T.*
Experimentelle Untersuchung zum Druckwiderstand der Alpen.
April 1988

Nr. 10: *Corsmeier, U.*
Analyse turbulenter Bewegungsvorgänge in der maritimen
atmosphärischen Grenzschicht.
Mai 1988

Nr. 11: *Walk, O. / Wieringa, J.(eds)*
Tsumeb Studies of the Tropical Boundary-Layer Climate.
Juli 1988

Nr. 12: *Degrazia, G. A.*
Anwendung von Ähnlichkeitsverfahren auf die turbulente Diffusion
in der konvektiven und stabilen Grenzschicht.
Januar 1989

Nr. 13: *Schädler, G.*
Numerische Simulationen zur Wechselwirkung zwischen Landober-
flächen und atmophärischer Grenzschicht.
November 1990

Nr. 14: *Heldt, K.*
Untersuchungen zur Überströmung eines mikroskaligen Hindernisses
in der Atmosphäre.
Juli 1991

Nr. 15: *Vogel, H.*
Verteilungen reaktiver Luftbeimengungen im Lee einer Stadt –
Numerische Untersuchungen der relevanten Prozesse.
Juli 1991

Nr. 16: *Höschele, K.(ed.)*
Planning Applications of Urban and Building Climatology – Proceedings
of the IFHP / CIB-Symposium Berlin, October 14-15, 1991.
März 1992

Nr. 17: *Frank, H. P.*
Grenzschichtstruktur in Fronten.
März 1992

Nr. 18: *Müller, A.*
Parallelisierung numerischer Verfahren zur Beschreibung von
Ausbreitungs- und chemischen Umwandlungsprozessen in der
atmosphärischen Grenzschicht.
Februar 1996

Nr. 19:   *Lenz, C.-J.*
Energieumsetzungen an der Erdoberfläche in gegliedertem Gelände.
Juni 1996

Nr. 20:   *Schwartz, A.*
Numerische Simulationen zur Massenbilanz chemisch reaktiver
Substanzen im mesoskaligen Bereich.
November 1996

Nr. 21:   *Beheng, K. D.*
Professor Dr. Franz Fiedler zum 60. Geburtstag.
Januar 1998

Nr. 22:   *Niemann, V.*
Numerische Simulation turbulenter Scherströmungen mit einem
Kaskadenmodell.
April 1998

Nr. 23:   *Koßmann, M.*
Einfluß orographisch induzierter Transportprozesse auf die Struktur
der atmosphärischen Grenzschicht und die Verteilung von
Spurengasen.
April 1998

Nr. 24:   *Baldauf, M.*
Die effektive Rauhigkeit über komplexem Gelände – Ein Störungs-
theoretischer Ansatz.
Juni 1998

Nr. 25:   *Noppel, H.*
Untersuchung des vertikalen Wärmetransports durch die Hangwind-
zirkulation auf regionaler Skala.
Dezember 1999

Nr. 26:   *Kuntze, K.*
Vertikaler Austausch und chemische Umwandlung von Spurenstoffen
über topographisch gegliedertem Gelände.
Oktober 2001

Nr. 27:   *Wilms-Grabe, W.*
Vierdimensionale Datenassimilation als Methode zur Kopplung zweier
verschiedenskaliger meteorologischer Modellsysteme.
Oktober 2001

Nr. 28: *Grabe, F.*
Simulation der Wechselwirkung zwischen Atmosphäre, Vegetation und Erdoberfläche bei Verwendung unterschiedlicher Parametrisierungsansätze.
Januar 2002

Nr. 29: *Riemer, N.*
Numerische Simulationen zur Wirkung des Aerosols auf die troposphärische Chemie und die Sichtweite.
Mai 2002

Nr. 30: *Braun, F. J.*
Mesoskalige Modellierung der Bodenhydrologie.
Dezember 2002

Nr. 31: *Kunz, M.*
Simulation von Starkniederschlägen mit langer Andauer über Mittelgebirgen.
März 2003

Nr. 32: *Bäumer, D.*
Transport und chemische Umwandlung von Luftschadstoffen im Nahbereich von Autobahnen – numerische Simulationen.
Juni 2003

Nr. 33: *Barthlott, C.*
Kohärente Wirbelstrukturen in der atmosphärischen Grenzschicht.
Juni 2003

Nr. 34: *Wieser, A.*
Messung turbulenter Spurengasflüsse vom Flugzeug aus.
Januar 2005

Nr. 35: *Blahak, U.*
Analyse des Extinktionseffektes bei Niederschlagsmessungen mit einem C-Band Radar anhand von Simulation und Messung.
Februar 2005

Nr. 36: *Bertram, I.*
Bestimmung der Wasser- und Eismasse hochreichender konvektiver Wolken anhand von Radardaten, Modellergebnissen und konzeptioneller Betrachtungen.
Mai 2005

Nr. 37: *Schmoeckel, J.*
Orographischer Einfluss auf die Strömung abgeleitet aus Sturmschäden im Schwarzwald während des Orkans „Lothar".
Mai 2006

Nr. 38:   *Schmitt, C.*
Interannual Variability in Antarctic Sea Ice Motion: Interannuelle
Variabilität antarktischer Meereis-Drift.
Mai 2006

Nr. 39:   *Hasel, M.*
Strukturmerkmale und Modelldarstellung der Konvektion über
Mittelgebirgen.
Juli 2006

Ab Band 40 erscheinen die Wissenschaftlichen Berichte des Instituts für Meteo-
rologie und Klimaforschung bei KIT Scientific Publishing (ISSN 0179-5619).
Die Bände sind unter www.ksp.kit.edu als PDF frei verfügbar oder als
Druckausgabe bestellbar.

Nr. 40:   *Lux, R.*
Modellsimulationen zur Strömungsverstärkung von orographischen
Grundstrukturen bei Sturmsituationen. (2007)
ISBN 978-3-86644-140-8

Nr. 41:   *Straub, W.*
Der Einfluss von Gebirgswellen auf die Initiierung und Entwicklung
konvektiver Wolken. (2008)
ISBN 978-3-86644-226-9

Nr. 42:   *Meißner, C.*
High-resolution sensitivity studies with the regional climate model
COSMO-CLM. (2008)
ISBN 978-3-86644-228-3

Nr. 43:   *Höpfner, M.*
Charakterisierung polarer stratosphärischer Wolken mittels hochauflö-
sender Infrarotspektroskopie. (2008)
ISBN 978-3-86644-294-8

Nr. 44:   *Rings, J.*
Monitoring the water content evolution of dikes. (2009)
ISBN 978-3-86644-321-1

Nr. 45:   *Riemer, M.*
Außertropische Umwandlung tropischer Wirbelstürme: Einfluss auf
das Strömungsmuster in den mittleren Breiten. (2012)
ISBN 978-3-86644-766-0

Nr. 46: *Anwender, D.*
Extratropical Transition in the Ensemble Prediction System of the ECMWF:
Case Studies and Experiments. (2012)
ISBN 978-3-86644-767-7

Nr. 47: *Rinke, R.*
Parametrisierung des Auswaschens von Aerosolpartikeln durch
Niederschlag. (2012)
ISBN 978-3-86644-768-4

Nr. 48: *Stanelle, T.*
Wechselwirkungen von Mineralstaubpartikeln mit thermodynamischen
und dynamischen Prozessen in der Atmosphäre über Westafrika. (2012)
ISBN 978-3-86644-769-1

Nr. 49: *Peters, T.*
Ableitung einer Beziehung zwischen der Radarreflektivität, der
Niederschlagsrate und weiteren aus Radardaten abgeleiteten
Parametern unter Verwendung von Methoden der multivariaten
Statistik. (2012)
ISBN 978-3-86644-323-5

Nr. 50: *Khodayar Pardo, S.*
High-resolution analysis of the initiation of deep convection forced
by boundary-layer processes. (2012)
ISBN 978-3-86644-770-7

Nr. 51: *Träumner, K.*
Einmischprozesse am Oberrand der konvektiven atmosphärischen
Grenzschicht. (2012)
ISBN 978-3-86644-771-4

Nr. 52: *Schwendike, J.*
Convection in an African Easterly Wave over West Africa and the
Eastern Atlantic: A Model Case Study of Hurricane Helene (2006)
and its Interaction with the Saharan Air Layer. (2012)
ISBN 978-3-86644-772-1

Nr. 53: *Lundgren, K.*
Direct Radiative Effects of Sea Salt on the Regional Scale. (2012)
ISBN 978-3-86644-773-8

Nr. 54: *Sasse, R.*
Analyse des regionalen atmosphärischen Wasserhaushalts unter
Verwendung von COSMO-Simulationen und GPS-Beobachtungen. (2012)
ISBN 978-3-86644-774-5

Nr. 55: *Grenzhäuser, J.*
Entwicklung neuartiger Mess- und Auswertungsstrategien für ein
scannendes Wolkenradar und deren Anwendungsbereiche. (2012)
ISBN 978-3-86644-775-2

Nr. 56: *Grams, C.*
Quantification of the downstream impact of extratropical transition
for Typhoon Jangmi and other case studies. (2013)
ISBN 978-3-86644-776-9

Nr. 57: *Keller, J.*
Diagnosing the Downstream Impact of Extratropical Transition
Using Multimodel Operational Ensemble Prediction Systems. (2013)
ISBN 978-3-86644-984-8